Mastering
The Art
of
War
by Sun Tzu

Mastering
The Art
of
War
by Sun Tzu

A questions and answers based approach of mastering
The Art of War by Sun Tzu

Titu Doley

This book is dedicated to Sun Tzu, who authored *The Art of War* more than 2000 years ago and Lionel Giles, based on whose English translation of *The Art of War* this book has been written.

CONTENTS

PART III
The Art of War (Without Commentary)

PREFACE

Sun Tzu, a great Chinese general and military strategist, wrote the *The Art of War* more than 2000 years ago. My first contact with a *The Art of War* book happened only a decade ago at a bookstore in Narita International Airport, Japan while waiting to catch a flight. It was an English translation of *The Art of War* with commentary. I bought a copy of the book to read later. While reading it for the first time I found it quite challenging to internalize the underlying message of Sun Tzu's teachings. Also, I had no way to validate if I had learnt Sun Tzu's teachings in the *The Art of War*. I searched the online books stores to find any book with exercises, quizzes or tests to check knowledge on Sun Tzu's teachings in the *The Art of War*. My search was futile, so I thought of creating such a book myself and hence this book. This book is written in the form of multiple choice questions and True or False questions to engage the reader to read, think and act by choosing the right answers. And in the process of answering the questions reader can also see for himself/herself if he/she thinks like Sun Tzu, the master military strategist.

I hope you find this book useful in creating competitive strategies, formulating tactics, excel in winning with ease and ultimately mastering the art of war.

INTRODUCTION

This book presents a questions and answers based approach of learning the teachings of Sun Tzu on art of warfare in his book *The Art of War*. The questions and answers have been compiled from one century old translation of Sun Tzu's *The Art of War* into English by Lionel Giles.

This book is organized into three parts -

Part I - In this part there are questions on each chapter of *The Art of War* and the correct answers are at the end of each chapter. It is recommended to first read either Part I or Part III of this book.

Part II - In this part the questions and answers on each chapter of *The Art of War* are together. The correct answers are in **bold** typeface and the wrong ones are in normal typeface. It is recommended to read the chapters in this part after reading the corresponding chapters in Part I.

Part III - This part contains *The Art of War* authored by Sun Tzu and translated into English by Lionel Giles.

Each part of this book has thirteen chapters, as in *The Art of War* by Sun Tzu. Here are some of the questions, chapter wise, readers will be able to answer by reading this book.

1. Laying Plans
- ❏ Why the art of war is of vital importance to the State?
- ❏ What are the five constant factors that govern the art of war?

❏ Which general will be victorious and who will fail?
❏ When seeking to determine the military conditions, what can be the basis of comparison?
❏ On what, all warfare are based?
❏ When does the general who win a battle make calculations in his temple?

2. Waging War
❏ In war what should be the great objective?
❏ Who knows the profitable way of carrying out war?
❏ What is the impact of prolonged warfare?
❏ How to augment one's own strength using the conquered foe?
❏ Why does a wise general make a point of foraging on the enemy?
❏ On whom depends, whether a nation shall be in peace or in peril?

3. Attack by Stratagem
❏ In practical art of war what is the best thing of all?
❏ What is supreme excellence in the art of war?
❏ What is the highest form of generalship?
❏ What does skillful leader do in warfare?
❏ What are the three ways in which a ruler can bring misfortune upon his army?
❏ What are the five essentials for victory in the art of war?

4. Tactical Dispositions
❏ When do the good fighters wait for an opportunity of defeating the enemy?
❏ Who provides the opportunity of defeating the enemy?
❏ How good fighters secure himself against defeat and make certain of defeating the enemy?
❏ When to use defensive and offensive tactics?
❏ When does a victorious strategist seek battle?
❏ What does the consummate leader cultivates and strictly adheres to?

5. Energy
❏ What is the principle for control of large force and a few men?
❏ How to fight with large army and small one under your command?
❏ Which methods to use for joining battle and for securing victory?
❏ Who sacrifices something that the enemy may snatch at it?

❏ What are the two methods of attack in battle?
❏ Why the clever combatant does not require too much energy from individuals?
❏ What are energy and decisions likened to?

6. Weak Points and Strong

❏ What can a clever combatant do to the enemy by holding out advantages?
❏ How can you be sure of succeeding in your attack?
❏ How can you ensure the safety of your defense?
❏ How can you force the enemy to engage in fight even though he be sheltered behind high rampart and deep ditch?
❏ Though the enemy be stronger in numbers, how we may prevent him from fighting?
❏ How would knowing the place and time of coming battle help you?
❏ What would happen if neither time nor place be known of the coming battle?
❏ In war what to avoid and what to strike at?

7. Maneuvering

❏ What must a general do before tactical maneuvering?
❏ What shows the knowledge of the artifice of deviation?
❏ With what type of army, maneuvering is dangerous?
❏ What should we know before entering into alliance with neighbors?
❏ Why practice dissimulation in war?
❏ How an army be formed so that it is impossible either for the brave to advance alone or the cowardly to retreat alone?
❏ When does a clever general avoid an army and when does he attacks it?
❏ What is art of studying moods, retaining self-possession, husbanding one's strength and studying circumstances?
❏ What not to do when you surround a desperate foe?

8. Variation in Tactics

❏ In war, from where the general receives his commands?
❏ What should the general know about tactics to handle his troops?
❏ What the student of war who is unversed in the art of varying his plans will fail at?
❏ In the wise leader's plans what considerations should be blended together?
❏ What does art of war teaches us to rely on?

❑ What are the besetting sins of a general, ruinous to the conduct of war?

9. The Army on the March
❑ Where to camp in mountain warfare?
❑ When to deliver attack and where to moor your craft in river warfare?
❑ Where to move to meet the enemy in river warfare?
❑ In crossing salt marshes what should be your sole concern?
❑ When the enemy keeps aloof and tries to provoke a battle, what is he anxious for?
❑ What is it when some soldiers are seen advancing and some retreating?
❑ If there is disturbance in the camp, what it tells about the general's authority?
❑ What does sight of men in the rank and file whispering together in small knot or speaking in subdued tones point to?
❑ What does peace proposals unaccompanied by a signed covenant indicate?
❑ When envoys are sent with compliments in their mouths, what is it sign of?
❑ What will happen if soldiers are punished before they have grown attached to you?
❑ How soldiers must be treated and kept under control?

10. Terrain
❑ What are the six kinds of terrain in art of war?
❑ What will ensue, if enemy on entangling ground is prepared for your coming, and you fail to defeat him?
❑ Why should you retreat on temporizing ground, even though the enemy offer an attractive bait?
❑ Should the enemy forestall you in occupying a narrow pass, when should you go after him and when you should not?
❑ The general who has attained a responsible post must study which six principles connected to Earth?
❑ What are the six calamities an army is exposed to, from faults for which the general is responsible?
❑ If fighting is sure to result in victory, then what you must do, even though the ruler forbid it?
❑ If fighting will not result in victory, then what you must not do, even at the ruler's bidding?

11. The Nine Situations

- ❑ What are the nine varieties of ground recognized in the art of war?
- ❑ Who could drive a wedge between the enemy's front and rear?
- ❑ When the enemy's men were united, who could managed to keep them in disorder?
- ❑ What will happen on seizing something which your opponent holds dear?
- ❑ What are the principles to be observed by an invading force?
- ❑ When do soldiers lose their sense of fear?
- ❑ Who conducts his army just as though he were leading a single man, willy-nilly by the hand?
- ❑ Whose business it is to be quiet and thus ensure secrecy; upright and just; and thus maintain order?
- ❑ At the critical moment, how does the leader of an army act?

12. The Attack by Fire

- ❑ What are the five ways of attacking with fire?
- ❑ What is the proper season for making attacks with fire?
- ❑ What are the special days for starting the conflagration?
- ❑ In attacking with fire, what are the five developments one should be prepared to meet?
- ❑ In attacking with fire, if there is an outbreak of fire but the enemy's soldiers remain quiet, what should you do?
- ❑ In every army, what are the five developments connected with fire must be known?
- ❑ Can an enemy be intercepted by means of water and robbed of all his belongings?

13. The Use of Spies

- ❑ What are the five classes of spies?
- ❑ Why use spies in war?
- ❑ What does foreknowledge enable the wise sovereign and the good general achieve things beyond the reach of the ordinary men?
- ❑ Why spies are most important element in war?
- ❑ Which class of spies be treated with the utmost liberty?

As a warm-up exercise, here is a Diagnostic Test with a set of 28 questions cherry picked from the thirteen chapters of the Part I. You may also take this Diagnostic Test to quickly check if you already know about Sun Tzu's teachings on art of warfare.

Diagnostic Test

Instructions:

1. Select the appropriate answer(s) from the questions with four options (a), (b), (c) and (d). In some questions there can be more than one correct answer.

2. In the questions with True or False options, choose the appropriate option.

Q1. Sun Tzu said: The art of war is of vital importance to the State. It is a matter of ___(i)___ and ___(ii)___, a road either to ___(iii)___ or to ___(iv)___. Hence it is a subject of inquiry which can on no account be neglected.

	(i)	(ii)	(iii)	(iv)
(a)	life	safety	death	ruin
(b)	safety	ruin	life	death
(c)	life	death	safety	ruin
(d)	life	ruin	safety	death

Q2. Sun Tzu said: The general who ___(i)___ a battle makes ___(ii)___ calculations in his temple before the battle is fought. The general who ___(iii)___ a battle makes but ___(iv)___ calculations beforehand.

	(i)	(ii)	(iii)	(iv)
(a)	wins	few	loses	many
(b)	loses	many	wins	few
(c)	wins	many	loses	few
(d)	loses	few	wins	many

Q3. Sun Tzu said: It is only one ___(i)___ is thoroughly ___(ii)___ with the ___(iii)___ of war that can thoroughly understand the ___(iv)___ way of carrying it on.

	(i)	(ii)	(iii)	(iv)
(a)	acquainted	profitable	who	evils
(b)	evils	profitable	acquainted	who
(c)	who	acquainted	evils	profitable
(d)	profitable	acquainted	who	evils

Q4. Sun Tzu said: The ___(i)___ of armies is the arbiter of the people's ___(ii)___, the man on whom it depends whether the ___(iii)___ shall be in ___(iv)___ or in peril.

(i)	(ii)	(iii)	(iv)
(a) fate	nation	peace	leader
(b) peace	rewards	nation	fate
(c) nation	fate	leader	peace
(d) leader	fate	nation	peace

Q5. Sun Tzu said: In the art of war, the supreme excellence is -

(a) to fight and conquer in all your battles.
(b) breaking the enemy's resistance without fighting.
(c) to fight and win no battle.
(d) to shatter and destroy the enemy's country.

Q6. Sun Tzu said: It is the rule in war, if our forces are equally matched in numbers to the enemy's, we can ___(i)___; if slightly inferior in numbers, we can ___(ii)___ the enemy; if quite unequal in every way, we can ___(iii)___ from the enemy.

(i)	(ii)	(iii)
(a) avoid	offer battle	flee
(b) avoid	flee	offer battle
(c) offer battle	avoid	flee
(d) flee	offer battle	avoid

Q7. Sun Tzu said: Security against defeat implies ___(i)___ tactics; ability to defeat the enemy means taking the ___(ii)___. Standing on the defensive indicates insufficient ___(iii)___; ___(iv)___, a superabundance of strength.

(i)	(ii)	(iii)	(iv)
(a) offensive	defensive	attacking	strength
(b) defensive	defensive	strength	attacking
(c) offensive	offensive	attacking	strength
(d) defensive	offensive	strength	attacking

Q8. Sun Tzu said: In war the ___(i)___ strategist only seeks battle ___(ii)___ the victory has been won, whereas he who is destined to ___(iii)___ ___(iv)___ fights and afterwards looks for victory.

(i)	(ii)	(iii)	(iv)
(a) defeat	before	victorious	after
(b) victorious	before	defeat	first
(c) defeat	after	victorious	first
(c) victorious	after	defeat	first

Q9. Sun Tzu said: The control of a large force is the same principle as the control of a few men: it is merely a question of -

(a) instituting signs and signals.
(b) direct and indirect maneuvers.
(c) science of weak points and strong.
(d) dividing up their numbers.

Q10. Sun Tzu said: In all fighting, the ___(i)___ method may be used for joining battle, but ___(ii)___ methods will be needed in order to secure victory.

(i)	(ii)
(a) indirect	direct
(b) direct	direct
(c) indirect	indirect
(d) direct	indirect

Q11. Sun Tzu said: Appear at points which the ___(i)___ must hasten to ___(ii)___ ; march swiftly to ___(iii)___ where you are ___(iv)___ .

(i)	(ii)	(iii)	(iv)
(a) not expected	places	defend	enemy
(b) enemy	places	defend	not expected
(c) not expected	defend	enemy	places
(d) enemy	defend	places	not expected

Q12. Sun Tzu said: If we wish to ___(i)___, the enemy can be ___(ii)___ to an engagement even though he be sheltered behind a high rampart and a deep ditch. All we need do is ___(iii)___ some other place that he will be obliged to ___(iv)___.

	(i)	(ii)	(iii)	(iv)
(a)	relieve	forced	fight	attack
(b)	fight	forced	attack	relieve
(c)	fight	relieve	attack	forced
(d)	attack	forced	fight	relieve

Q13. Sun Tzu said: If enemy sends reinforcements everywhere, he will everywhere be strong. (True or False)

Q14. Sun Tzu said: It is a military axiom not to ___(i)___ against the enemy, nor to ___(ii)___ him when he comes ___(iii)___.

	(i)	(ii)	(iii)
(a)	oppose	downhill	advance uphill
(b)	advance uphill	oppose	downhill
(c)	advance uphill	downhill	advance uphill
(d)	downhill	oppose	advance uphill

Q15. Sun Tzu said: When you ___(i)___ an army, ___(ii)___ an outlet free. Do not ___(iii)___ a desperate foe too hard. Such is the art of warfare.

	(i)	(ii)	(iii)
(a)	leave	press	surround
(b)	press	surround	leave
(c)	surround	press	leave
(d)	surround	leave	press

Q16. Sun Tzu said: There are roads which must not be ___(i)___, armies which must be not ___(ii)___, towns which must be ___(iii)___, positions which must not be ___(iv)___, commands of the sovereign which must not be obeyed.

	(i)	(ii)	(iii)	(iv)
(a)	besieged	contested	attacked	followed
(b)	followed	contest	attacked	besieged
(c)	besieged	contested	followed	attacked
(d)	followed	attacked	besieged	contested

Q17. Sun Tzu said: The art of war teaches us to rely not on the likelihood of the enemy's ___(i)___, but on our own ___(ii)___ to receive him; not on the chance of his ___(iii)___, but rather on the fact that we have made our position ___(iv)___.

	(i)	(ii)	(iii)	(iv)
(a)	readiness	unassailable	not coming	not attacking
(b)	unassailable	readiness	not attacking	not coming
(c)	not coming	unassailable	not attacking	readiness
(d)	not coming	readiness	not attacking	unassailable

Q18. Which of the following is (are) true of river warfare?

(a) When an invading force crosses a river in its onward march, do not advance to meet it in mid-stream. It will be best to let half the army get across, and then deliver your attack.

(b) After crossing a river, you should get far away from it.

(c) If you are anxious to fight, you should not go to meet the invader near a river which he has to cross.

(d) Moor your craft higher up than the enemy, and facing the sun. Do not move down-stream to meet the enemy.

Q19. Sun Tzu said: When ___(i)___ are sent with ___(ii)___ in their mouths, it is a sign that the ___(iii)___ wishes for a ___(iv)___.

	(i)	(ii)	(iii)	(iv)
(a)	truce	enemy	envoys	compliments
(b)	enemy	truce	envoys	compliments
(c)	envoys	compliments	enemy	truce
(d)	compliments	truce	envoys	enemy

Q20. If fighting is sure to result in victory, then you ___(i)___ fight, even though the ruler ___(ii)___ it; if fighting will not result in victory, then you ___(iii)___ fight even at the ruler's ___(iv)___ .

(i)	(ii)	(iii)	(iv)
(a) must not	forbid	must	bidding
(b) must not	bidding	must	forbid
(c) must	forbid	must not	bidding
(d) forbid	must	must not	bidding

Q21. If we ___(i)___ that our own men are in a ___(ii)___ to attack, but are ___(iii)___ that the enemy is ___(iv)___ to attack, we have gone only halfway towards victory.

(i)	(ii)	(iii)	(iv)
(a) condition	not open	unaware	know
(b) unaware	not open	condition	know
(c) condition	unaware	know	not open
(d) know	condition	unaware	not open

Q22. Sun Tzu said: ___(i)___ is the essence of war: take advantage of the enemy's ___(ii)___ , make your way by ___(iii)___ routes, and attack ___(iv)___ spots.

(i)	(ii)	(iii)	(iv)
(a) unexpected	rapidity	unguarded	unreadiness
(b) unguarded	unreadiness	rapidity	unexpected
(c) rapidity	unreadiness	unexpected	unguarded
(d) unexpected	unreadiness	rapidity	unguarded

Q23. Sun Tzu said: Success in warfare is gained by carefully accommodating ourselves to the enemy's purpose. (True or False)

Q24. Sun Tzu said: At first, then, exhibit the coyness of a maiden, until the ___(i)___ gives you an ___(ii)___; afterwards emulate the rapidity of a running hare, and it will be too ___(iii)___ for the enemy to ___(iv)___ you.

	(i)	(ii)	(iii)	(iv)
(a)	late	enemy	oppose	opening
(b)	enemy	late	opening	oppose
(c)	opening	enemy	oppose	late
(d)	enemy	opening	late	oppose

Q25. Sun Tzu said: Those who use ___(i)___ as an aid to the attack show ___(ii)___; those who use ___(iii)___ as an aid to the attack gain an accession of ___(iv)___.

	(i)	(ii)	(iii)	(iv)
(a)	intelligence	strength	water	fire
(b)	water	intelligence	strength	fire
(c)	fire	intelligence	water	strength
(d)	strength	fire	intelligence	water

Q26. Sun Tzu said: __(i)__ not unless you see an __(ii)__; use not your troops unless there is something to be gained; __(iii)__ not unless the position is __(iv)__.

	(i)	(ii)	(iii)	(iv)
(a)	fight	critical	move	advantage
(b)	advantage	critical	move	fight
(c)	move	advantage	fight	critical
(d)	critical	fight	move	advantage

Q27. Sun Tzu said: The five classes of spies are : (1) Local spies; (2) __(i)__; (3) converted spies; (4) __(ii)__; (5) surviving spies.

	(i)	(ii)
(a)	outward spies	inverted spies
(b)	inward spies	outward spies
(c)	doomed spies	inverted spies
(d)	inward spies	doomed spies

Q28. Sun Tzu said: It is essential that the converted spy be treated with the utmost liberality. Why ?

(a) It is through the information brought by the converted spy that we are able to acquire and employ local and inward spies.

(b) It is owing to the information brought by the converted spy, that we can cause the doomed spy to carry false tidings to the enemy.

(c) It is by information brought by the converted spy that the surviving spy can be used on appointed occasions.

(d) The end and aim of spying in all its five varieties is knowledge of the enemy; and this knowledge can only be derived, in the first instance, from the converted spy.

Diagnostic Test Answers

Q1. (c)
Q2. (c)
Q3. (c)
Q4. (d)
Q5. (b)
Q6. (c)
Q7. (d)
Q8. (d)
Q9. (d)
Q10. (d)
Q11. (d)
Q12. (b)
Q13. True
Q14. (b)
Q15. (d)
Q16. (d)
Q17. (d)
Q18. (a), (b), (c)
Q19. (c)
Q20. (c)
Q21. (d)
Q22. (c)
Q23. True
Q24. (d)
Q25. (c)
Q26. (c)
Q27. (d)
Q28. (a), (b), (c), (d)

PART I

In this part there are questions on each chapter of *The Art of War* and the correct answers at the end of each chapter.

1
LAYING PLANS

Instructions:
Select the appropriate answer(s) from the questions with four options (a),
(b), (c) and (d). In some questions there can be more than one correct
answer.

Q1. Sun Tzu said: The art of war is of vital importance to the State. It is a matter of ___(i)___ and ___(ii)___, a road either to ___(iii)___ or to ___(iv)___. Hence it is a subject of inquiry which can on no account be neglected.

(i)	(ii)	(iii)	(iv)
(a) life	safety	death	ruin
(b) safety	ruin	life	death
(c) life	death	safety	ruin
(c) life	ruin	safety	death

Q2. What are the five constant factors that govern the art of war ?

(a) The Moral Law; Heaven; Sea; The Commander; Method and Discipline
(b) The Moral Law; Heaven; Earth; The Commander; The State
(c) The Moral Law; Heaven; Earth; The Commander; Method and Discipline
(d) The Moral Law; Heaven; Earth; Sea; The Commander

Q3. What does the Moral Law, a constant factor that governs the art of war, causes the people to be?

17

(a) It causes the people to be in complete discord with their enemy.
(b) It causes the people to be in complete accord with their ruler.
(c) It causes the ruler to be in complete accord with the people.
(d) It causes the people to be in complete accord with themselves.

Q4. What does Heaven, one of the five constant factors that govern the art of war, do not signify?

(a) day and night
(b) cold and heat
(c) danger and security
(d) times and seasons

Q5. What does the Earth, one of the five constant factors that govern the art of war, comprises of?

(a) danger and security
(b) distances great and small
(c) open ground and narrow passes
(d) the chances of life and death

Q6. The Commander, a constant factor that governs the art of war, stands for virtues of -

(a) wisdom and sincerity only
(b) wisdom, sincerity and benevolence only
(c) courage and strictness only
(d) wisdom, sincerity, benevolence, courage and strictness

Q7. Sun Tzu said: By Method and Discipline are to be understood the marshaling of the army in its proper ___(i)___, the graduations of rank among the ___(ii)___, the ___(iii)___ of roads by which supplies may reach the army, and the control of military ___(iv)___.

	(i)	(ii)	(iii)	(iv)
(a)	officers	subdivisions	expenditure	maintenance
(b)	expenditure	subdivisions	maintenance	officers
(c)	maintenance	expenditure	subdivisions	officers
(d)	subdivisions	officers	maintenance	expenditure

Q8. Sun Tzu said: The general who knows the ___(i)___ constant factors which govern the art of war, will be ___(ii)___; he who knows them not will ___(iii)___.

	(i)	(ii)	(iii)
(a)	three	victorious	fail
(b)	four	fail	victorious
(c)	five	victorious	fail
(d)	six	fail	victorious

Q9. In your deliberations, when seeking to determine the military conditions, which of the following can be the basis of a comparison?

(a) Which of the two sovereigns is imbued with the Moral law?
(b) Which of the two generals has most ability?
(c) With whom lie the advantages derived from Heaven and Earth?
(d) On which side is discipline most rigorously enforced?

Q10. By means of which considerations, Sun Tzu can forecast victory or defeat?

(a) (1) Which of the two sovereigns is imbued with the Moral law?
    (2) Which of the two generals has most ability?
(b) (1) With whom lie the advantages derived from Heaven and Earth?
    (2) On which side is discipline most rigorously enforced?
(c) (1) Which army is stronger?
    (2) On which side are officers and men more highly trained?
(d) In which army is there the greater constancy both in reward and punishment?

Q11. Sun Tzu said: The general that hearkens to my counsel and acts upon it, will ___(i)___ :- let such a one be ___(ii)___ in command! The general that hearkens not to my counsel nor acts upon it, will suffer ___(iii)___ :- let such a one be ___(iv)___ !

	(i)	(ii)	(iii)	(iv)
(a)	conquer	dismissed	defeat	retained
(b)	defeat	retained	conquer	dismissed
(c)	conquer	retained	defeat	dismissed
(d)	defeat	conquer	dismissed	retained

Q12. Sun Tzu said: While heading the profit of my counsel, avail yourself also of any helpful ___(i)___ over and beyond the ___(ii)___ rules. According as circumstances are ___(iii)___, one should modify one's ___(iv)___.

	(i)	(ii)	(iii)	(iv)
(a)	plans	favorable	ordinary	circumstances
(b)	ordinary	favorable	plans	circumstances
(c)	circumstances	ordinary	favorable	plans
(d)	plans	circumstances	ordinary	favorable

13. Sun Tzu said: All warfare is based on deception. Hence -

(a) when able to attack, we must seem unable.
(b) when using our forces, we must seem inactive.
(c) when we are near, we must make the enemy believe we are far away.
(d) when far away, we must make enemy believe we are near.

Q14. Sun Tzu said: These military devices, leading to victory, must not be divulged beforehand -

(a) Hold out baits to entice the enemy. Feign disorder, and crush him.
(b) If he is not secure at all points, be prepared for him. If he is in superior strength, evade him.

(c) If he is taking his ease, give him no rest. If his forces are united, separate them.

(d) Attack enemy where he is unprepared, appear where you are expected.

Q15. Sun Tzu said: If your opponent is of choleric temper, seek to ___(i)___ him. Pretend to be ___(ii)___, that he may grow ___(iii)___.

(i)	(ii)	(iii)
(a) weak	arrogant	irritate
(b) irritate	arrogant	weak
(c) weak	irritate	weak
(d) irritate	weak	arrogant

Q16. Sun Tzu said: The general who ___(i)___ a battle makes ___(ii)___ calculations in his temple before the battle is fought. The general who ___(iii)___ a battle makes but ___(iv)___ calculations beforehand.

(i)	(ii)	(iii)	(iv)
(a) wins	few	loses	many
(b) loses	many	wins	few
(c) wins	many	loses	few
(d) loses	few	wins	many

1
LAYING PLANS

Answers

Q1. (c)
Q2. (c)
Q3. (b)
Q4. (c)
Q5. (a), (b), (c), (d)
Q6. (d)
Q7. (d)
Q8. (c)
Q9. (a), (b), (c), (d)
Q10. (a), (b), (c), (d)
Q11. (c)
Q12. (c)
Q13. (a), (b), (c), (d)
Q14. (a), (c)
Q15. (d)
Q16. (c)

2

WAGING WAR

Instructions:
1. Select the appropriate answer(s) from the questions with four options (a), (b), (c) and (d). In some questions there can be more than one correct answer.
2. In the questions with True or False options, choose the appropriate option.

Q1. Sun Tzu said: When you engage in actual fighting, if ___(i)___ is long in coming, then men's weapons will grow ___(ii)___ and their ardor will be ___(iii)___. If you lay siege to a town, you will ___(iv)___ your strength.

(i)	(ii)	(iii)	(iv)
(a) victory	exhaust	damped	dull
(b) exhaust	dull	victory	damped
(c) damped	exhaust	dull	victory
(d) victory	dull	damped	exhaust

Q2. Sun Tzu said: When you engage in actual fighting, if the campaign is ___(i)___, the resources of the ___(ii)___ will not be equal to the ___(iii)___.

(i)	(ii)	(iii)
(a) strain	State	protracted
(b) State	protracted	strain
(c) protracted	State	strain
(d) strain	protracted	State

Q3. Sun Tzu said: In a prolonged warfare, when your weapons are dulled, your ardor damped, your strength exhausted and your treasure spent, other chieftains will spring up to take advantage of your extremity. Then no man, however wise, will be able to avert the consequences that must ensue. (True or False)

Q4. Sun Tzu said: There is no instance of a country having benefited from prolonged warfare. (True or False)

Q5. Sun Tzu said: It is only one ___(i)___ is thoroughly ___(ii)___ with the ___(iii)___ of war that can thoroughly understand the ___(iv)___ way of carrying it on.

(i)	(ii)	(iii)	(iv)
(a) acquainted	profitable	who	evils
(b) evils	profitable	acquainted	who
(c) who	acquainted	evils	profitable
(d) profitable	acquainted	who	evils

Q6. Sun Tzu said: The skillful soldier does not raise a second levy, neither are his supply-wagons loaded more than twice. (True or False)

Q7. Sun Tzu said: Bring war material with you from home, but forage on the enemy. Thus the army will have food enough for its needs. (True or False)

Q8. A wise general makes a point of foraging on the enemy, why?

(a) Contributing to maintain an army at a distance causes the people to be impoverished.
(b) The proximity of an army causes prices to go up; and high prices cause the people's substance to be drained away.
(c) When people's substance is drained away, the peasantry will be afflicted by heavy exactions.
(d) One cartload of the enemy's provisions is equivalent to twenty of one's own.

Q9. Sun Tzu said: In order to kill the ___(i)___, our men must be roused to ___(ii)___; that there may be ___(iii)___ from defeating the enemy, they must have their ___(iv)___.

(i)	(ii)	(iii)	(iv)
(a) anger	enemy	rewards	advantage

(b) advantage   rewards   anger   enemy
(c) enemy   anger   advantage   rewards
(d) rewards   enemy   anger   advantage

Q10. Which of the following is (are) using the conquered foe to augment one's own strength ?

(a) The captured soldiers kindly treated and kept.
(b) Our own flags substituted for those of the enemy on chariots taken in chariot fighting, and the chariots mingled and used in conjunction with ours.
(c) Bringing war material with you from home, but foraging on the enemy.
(d) None of these.

Q11. Sun Tzu said: In war let your great object be victory, not lengthy campaigns. (True or False)

Q12. Sun Tzu said: The ___(i)___ of armies is the arbiter of the people's ___(ii)___, the man on whom it depends whether the ___(iii)___ shall be in ___(iv)___ or in peril.

(i)	(ii)	(iii)	(iv)
(a) fate	nation	peace	leader
(b) peace	rewards	nation	fate
(c) nation	fate	leader	peace
(d) leader	fate	nation	peace

2
WAGING WAR

Answers

Q1. (d)
Q2. (c)
Q3. True
Q4. True
Q5. (c)
Q6. True
Q7. True
Q8. (a), (b), (c), (d)
Q9. (c)
Q10. (a), (b), (c)
Q11. True
Q12. (d)

3
ATTACK BY STRATAGEM

Instructions:
1. Select the appropriate answer(s) from the questions with four options (a), (b), (c) and (d). In some questions there can be more than one correct answer.
2. In the questions with True or False options, choose the appropriate option.

Q1. Sun Tzu said: In the practical art of war, the best thing of all is -

(a) to shatter and destroy the enemy's country.
(b) to recapture an army entire than to destroy it.
(c) to take the enemy's country whole and intact.
(d) to capture a regiment, a detachment or a company entire than to destroy them.

Q2. Sun Tzu said: In the art of war, the supreme excellence is –

(a) to fight and conquer in all your battles.
(b) breaking the enemy's resistance without fighting.
(c) to fight and win no battle.
(d) to shatter and destroy the enemy's country.

Q3. Arrange the following four forms of generalship from the highest form to the lowest form.

(I) to attack the enemy's army in the field
(II) to besiege walled cities
(III) to balk the enemy's plans
(IV) to prevent the junction of the enemy's forces

27

(a) (I), (II), (III), (IV)
(b) (III), (I), (II), (IV)
(c) (II), (IV), (I), (III)
(d) (III), (IV), (I), (II)

Q4. Sun Tzu said: The rule is, not to besiege walled cities if it can possibly be avoided. Why?

(a) The preparation of mantlets, movable shelters, and various implements of war, will take up three whole months.
(b) The piling up of mounds over against the walls will take three months.
(c) The general, unable to control his irritation, will launch his men to the assault like swarming ants, with the result that one-third of his men are slain, while the town still remains untaken. Such are the disastrous effects of a siege.
(d) None of these.

Q5. In war, what does skillful leader do ?

(a) The skillful leader subdues the enemy's troops without any fighting.
(b) He captures enemy cities without laying siege to them.
(c) He overthrows enemy kingdom without lengthy operations in the field.
(d) With his forces intact he disputes the mastery of the Empire, and thus, without losing a man, he completes triumph.

Q6. Sun Tzu said: It is the rule in war, if our forces are ten to the enemy's one, then

(a) attack the enemy
(b) offer battle to the enemy
(c) surround the enemy
(d) avoid the enemy

Q7. Sun Tzu said: It is the rule in war, if our forces are five to the enemy's one then attack the enemy. (True or False)

Q8. Sun Tzu said: It is the rule in war, if our forces are twice as numerous as the enemy, then

(a) attack the enemy

(b) offer battle to the enemy
(c) avoid the enemy
(d) divide our army into two

Q9. Sun Tzu said: It is the rule in war, if our forces are equally matched in numbers to the enemy's, we can ___(i)___; if slightly inferior in numbers, we can ___(ii)___ the enemy; if quite unequal in every way, we can ___(iii)___ from the enemy.

(i)	(ii)	(iii)
(a) avoid	offer battle	flee
(b) avoid	flee	offer battle
(c) offer battle	avoid	flee
(d) flee	offer battle	avoid

10. Sun Tzu said: The ___(i)___ is the bulwark of the ___(ii)___; if the bulwark is complete at all points; the State will be ___(iii)___; if the bulwark is defective, the State will be ___(iv)___.

(i)	(ii)	(iii)	(iv)
(a) State	strong	general	weak
(b) strong	State	weak	general
(c) general	State	strong	weak
(d) State	general	weak	strong

Q11. The way(s) in which a ruler can bring misfortune upon his army is (are) -

(a) By commanding the army to advance or to retreat, being ignorant of the fact that it cannot obey.
(b) By attempting to govern an army in the same way as he administers a kingdom, being ignorant of the conditions which obtain in an army.
(c) By employing the officers of his army without discrimination, through ignorance of the military principle of adaptation to circumstances.
(d) None of these.

Q12. In art of war, which of the following is (are) part of the five essentials for victory?

(a) He will win who knows when to fight and when not to fight.

(b) He will win who knows how to handle both superior and inferior forces.
(c) He will win who, prepared himself, waits to take the enemy unprepared.
(d) He will win whose army is animated by the same spirit throughout all its ranks.

Q13. Sun Tzu said: If you know the enemy and know yourself then

(a) for every victory gained you will also suffer a defeat.
(b) you will succumb in every battle.
(c) you need not fear the result of a hundred battles.
(d) you can't break the enemy's resistance without fighting.

Q14. Sun Tzu said: If you know yourself but not the enemy then

(a) you will succumb in every battle.
(b) you need not fear the result of a hundred battles.
(c) for every victory gained you will also suffer a defeat.
(d) you will break the enemy's resistance without fighting.

Q15. Sun Tzu said: If you know neither the enemy nor yourself then

(a) you need not fear the result of a hundred battles.
(b) you will break the enemy's resistance without fighting.
(c) for every victory gained you will also suffer a defeat.
(d) you will succumb in every battle.

3
ATTACK BY STRATAGEM

Answers

Q1. (c)
Q2. (b)
Q3. (d)
Q4. (a), (b), (c)
Q5. (a), (b), (c), (d)
Q6. (c)
Q7. True
Q8. (d)
Q9. (c)
Q10. (c)
Q11. (a), (b), (c)
Q12. (a), (b), (c), (d)
Q13. (c)
Q14. (c)
Q15. (d)

4
TACTICAL DISPOSITIONS

Instructions:
1. Select the appropriate answer(s) from the questions with four options (a), (b), (c) and (d). In some questions there can be more than one correct answer.
2. In the questions with True or False options, choose the appropriate option.

Q1. Sun Tzu said: The good ___(i)___ of old first put themselves beyond the possibility of ___(ii)___, and then ___(iii)___ for an opportunity of defeating the ___(iv)___.

(i)	(ii)	(iii)	(iv)
(a) enemy	fighters	defeat	waited
(b) fighters	enemy	waited	defeat
(c) waited	fighters	enemy	defeat
(d) fighters	defeat	waited	enemy

Q2. Sun Tzu said: To secure ourselves against defeat lies in our own hands, but -

(a) the opportunity of defeating the enemy is provided by ourselves.
(b) the opportunity of defeating the enemy is not provided by the enemy himself.
(c) the opportunity of defeating the enemy is provided by the enemy himself.
(d) the opportunity of defeating ourselves is provided by the enemy.

Q3. Sun Tzu said: The good fighter is able to secure himself against defeat, but cannot make certain of defeating the enemy. (True or False)

Q4. Sun Tzu said: Security against defeat implies ___(i)___ tactics; ability to defeat the enemy means taking the ___(ii)___ . Standing on the defensive indicates insufficient ___(iii)___ ; ___(iv)___ , a superabundance of strength.

(i)	(ii)	(iii)	(iv)
(a) offensive	defensive	attacking	strength
(b) defensive	defensive	strength	attacking
(c) offensive	offensive	attacking	strength
(d) defensive	offensive	strength	attacking

Q5. Sun Tzu said: The general who is skilled in ___(i)___ hides in the most secret recesses of the earth; he who is skilled in ___(ii)___ flashes forth from the topmost heights of heaven. Thus on the one hand we have ability to ___(iii)___ ourselves; on the other, a ___(iv)___ that is complete.

(i)	(ii)	(iii)	(iv)
(a) attack	defense	protect	victory
(b) victory	attack	protect	defense
(c) protect	defense	victory	attack
(d) defense	attack	protect	victory

Q6. Sun Tzu said: To see victory only when it is within the ken of the common herd is the acme of excellence. (True or False)

Q7. Sun Tzu said: It is not the acme of excellence if you fight and conquer and the whole Empire says, "Well done!" (True or False)

Q8. Sun Tzu said: What the ancients called a clever fighter is the one -

(a) who wins his battles by making no mistakes.
(b) who not only wins, but excels in winning with ease.
(c) whose victories bring him neither reputation for wisdom nor credit for courage.

(d) who puts himself into a position which makes defeat impossible, and does not miss the moment for defeating the enemy.

Q9. Sun Tzu said: In war the ___(i)___ strategist only seeks battle ___(ii)___ the victory has been won, whereas he who is destined to ___(iii)___ ___(iv)___ fights and afterwards looks for victory.

(i)	(ii)	(iii)	(iv)
(a) defeat	before	victorious	after
(b) victorious	before	defeat	first
(c) defeat	after	victorious	first
(d) victorious	after	defeat	first

Q10. Sun Tzu said: The consummate ___(i)___ cultivates the ___(ii)___, and strictly adheres to method and ___(iii)___; thus it is in his power to ___(iv)___ success.

(i)	(ii)	(iii)	(iv)
(a) discipline	control	moral law	leader
(b) moral law	discipline	leader	control
(c) leader	moral law	discipline	control
(d) control	leader	moral law	discipline

Q11. Sun Tzu said: In respect of military method, we have, firstly, Measurement; secondly, Estimation of quantity; thirdly, Calculation; fourthly, Balancing of chances; fifthly, Victory. Measurement owes its existence to ___(i)___; Estimation of quantity to ___(ii)___; Calculation to Estimation of quantity; Balancing of chances to ___(iii)___; and ___(iv)___ to Balancing of chances.

(i)	(ii)	(iii)	(iv)
(a) Victory	Calculation	Measurement	Earth
(b) Calculation	Victory	Earth	Measurement
(c) Earth	Measurement	Calculation	Victory
(d) Measurement	Calculation	Victory	Earth

Q12. Sun Tzu said: A ___(i)___ army opposed to a ___(ii)___ one, is as a pound's weight placed in the scale against a single grain. The ___(iii)___ of a conquering ___(iv)___ is like the bursting of pent-up waters into a chasm a thousand fathoms deep.

(i)	(ii)	(iii)	(iv)
(a) routed	onrush	force	victorious
(b) onrush	victorious	routed	victorious
(c) victorious	routed	onrush	force
(d) routed	victorious	force	onrush

4

TACTICAL DISPOSITIONS

Answers

Q1. (d)
Q2. (c)
Q3. True
Q4. (d)
Q5. (d)
Q6. False
Q7. True
Q8. (a), (b), (c), (d)
Q9. (d)
Q10. (c)
Q11. (c)
Q12. (c)

5
ENERGY

Instructions:
1. Select the appropriate answer(s) from the questions with four options (a),
(b), (c) and (d). In some questions there can be more than one correct
answer.
2. In the questions with True or False options, choose the appropriate option.

Q1. Sun Tzu said: The control of a large force is the same principle as the control of a few men: it is merely a question of -

(a) instituting signs and signals.
(b) direct and indirect maneuvers.
(c) science of weak points and strong.
(d) dividing up their numbers.

Q2. Sun Tzu said: Fighting with a large army under your command is nowise different from fighting with a small one: it is merely a question of -

(a) science of weak points and strong.
(b) dividing up their numbers.
(c) instituting signs and signals.
(d) direct and indirect maneuvers.

Q3. Sun Tzu said: To ensure that your army may withstand the brunt of the enemy's attack and remain unshaken - this is effected by

(a) science of weak points and strong.

(b) instituting signs and signals.
(c) dividing up their numbers.
(d) direct and indirect maneuvers.

Q4. Sun Tzu said: That the impact of your army may be like a grindstone dashed against an egg - this is effected by

(a) instituting signs and signals.
(b) direct and indirect maneuvers.
(c) science of weak points and strong.
(d) Heaven and Earth

Q5. Sun Tzu said: In all fighting, the ___(i)___ method may be used for joining battle, but ___(ii)___ methods will be needed in order to secure victory.

(i)	(ii)
(a) indirect	direct
(b) direct	direct
(c) indirect	indirect
(d) direct	indirect

Q6. Sun Tzu said: _____, efficiently applied, are inexhaustible as Heaven and Earth, unending as the flow of rivers and streams; like the sun and moon, they end but to begin anew; like the four seasons, they pass away to return once more.

(a) Direct tactics
(b) Indirect tactics
(c) Direct and indirect tactics
(d) Science of weak points

Q7. Sun Tzu said: In battle, there are not more than two methods of attack - the ___(i)___ and the ___(ii)___; yet these two in combination give rise to an ___(iii)___ series of maneuvers.

(i)	(ii)	(iii)
(a) direct	endless	indirect
(b) endless	indirect	direct

(c) direct   indirect   endless

(d) indirect   endless   direct

Q8. Sun Tzu said: The direct and the indirect methods of attack lead on to each other in turn. It is like moving in a circle - you never come to an end. (True or False)

Q9. Sun Tzu said: The ___(i)___ of troops is like the rush of a torrent which will even roll stones along in its course. The ___(ii)___ of decision is like the well-timed swoop of a falcon which enables it to strike and destroy its victim. Therefore the good fighter will be ___(iii)___ in his onset, and ___(iv)___ in his decision.

(i)	(ii)	(iii)	(iv)
(a) quality	onset	prompt	terrible
(b) quality	onset	terrible	prompt
(c) onset	quality	terrible	prompt
(d) onset	quality	prompt	terrible

Q10. Sun Tzu said: ___(i)___ may be likened to the ___(ii)___ of a crossbow; ___(iii)___, to the ___(iv)___ of a trigger.

(i)	(ii)	(iii)	(iv)
(a) decision	releasing	energy	bending
(b) decision	bending	energy	releasing
(c) energy	releasing	decision	bending
(d) energy	bending	decision	releasing

Q11. Sun Tzu said: Amid the turmoil and tumult of ___(i)___, there may be seeming disorder and yet no real ___(ii)___ at all; amid confusion and chaos, your array may be without ___(iii)___, yet it will be proof against ___(iv)___.

(i)	(ii)	(iii)	(iv)
(a) defeat	head or tail	disorder	battle
(b) disorder	defeat	head or tail	battle
(c) battle	disorder	head or tail	defeat
(d) head or tail	battle	defeat	disorder

Q12. Sun Tzu said: Hide order beneath the cloak of disorder. Conceal courage under a show of timidity. (True or False)

Q13. Sun Tzu said: One who is skillful at keeping the enemy on the move maintains ___(i)___ appearances, according to which the enemy will act. He ___(ii)___ something, that the enemy may snatch at it. By holding out baits, he keeps him on the ___(iii)___; then with a body of picked men he lies in ___(iv)___ for him.

(i)	(ii)	(iii)	(iv)
(a) march	wait	deceitful	sacrifices
(b) wait	march	sacrifices	deceitful
(c) deceitful	sacrifices	march	wait
(d) sacrifices	wait	march	deceitful

Q14. Sun Tzu said: The clever combatant looks to the effect of ___(i)___ energy, and does not require too much from ___(ii)___. Hence his ability to pick out the right ___(iii)___ and utilize combined ___(iv)___.

(i)	(ii)	(iii)	(iv)
(a) individuals	combined	energy	men
(b) men	individuals	combined	energy
(c) individuals	energy	combined	men
(d) combined	individuals	men	energy

Q15. Sun Tzu said: The ___(i)___ developed by good fighting men is as the ___(ii)___ of a round stone rolled down a mountain thousands of feet in height.

(i)	(ii)
(a) momentum	energy
(b) energy	energy
(c) momentum	momentum
(d) energy	momentum

5
ENERGY

Answers

Q1. (d)
Q2. (c)
Q3. (d)
Q4. (c)
Q5. (d)
Q6. (b)
Q7. (c)
Q8. True
Q9. (c)
Q10. (d)
Q11. (c)
Q12. True
Q13. (c)
Q14. (d)
Q15. (d)

6
WEAK POINTS AND STRONG

Instructions:
1. Select the appropriate answer(s) from the questions with four options (a), (b), (c) and (d). In some questions there can be more than one correct answer.
2. In the questions with True or False options, choose the appropriate option.

Q1. Sun Tzu said: Whoever is first in the field and ___(i)___ the coming of the enemy, will be ___(ii)___ for the fight; whoever is ___(iii)___ in the field and has to hasten to battle will arrive ___(iv)___ .

(i)	(ii)	(iii)	(iv)
(a) second	exhausted	fresh	awaits
(b) awaits	exhausted	second	fresh
(c) exhausted	awaits	fresh	second
(d) awaits	fresh	second	exhausted

Q2. Sun Tzu said: The clever combatant imposes his will on the enemy, but does not allow the enemy's will to be imposed on him. (True or False)

Q3. What can a clever combatant do by holding out advantages to him?

(a) He can cause the enemy to approach of his own accord
(b) By inflicting damage, he can make it impossible for the enemy to draw near.
(c) If the enemy is taking his ease, he can harass him; if well supplied with food, he can starve him out.

(d) If enemy is quietly encamped, he can force enemy to move.

Q4. Sun Tzu said: Appear at points which the ___(i)___ must hasten to ___(ii)___; march swiftly to ___(iii)___ where you are ___(iv)___.

(i)	(ii)	(iii)	(iv)
(a) not expected	places	defend	enemy
(b) enemy	places	defend	not expected
(c) not expected	defend	enemy	places
(d) enemy	defend	places	not expected

Q5. Sun Tzu said: An army may march great distances without distress, if it marches through country where the enemy is not. (True or False)

Q6. Sun Tzu said: You can be sure of succeeding in your ___(i)___ if you only attack places which are ___(ii)___. You can ensure the safety of your ___(iii)___ if you only hold positions that cannot be ___(iv)___.

(i)	(ii)	(iii)	(iv)
(a) defense	undefended	attacks	attacked
(b) defense	attacked	attacks	undefended
(c) attacks	undefended	defense	attacked
(d) attacks	attacked	defense	undefended

Q7. Sun Tzu said: A general is skillful in attack whose opponent does not know what to ___(i)___; and he is skillful in ___(ii)___ whose opponent does not know what to ___(iii)___.

(i)	(ii)	(iii)
(a) attack	defense	defend
(b) defend	attack	defense
(c) attack	defend	defense
(d) defend	defense	attack

Q8. Sun Tzu said: You may ___(i)___ and be absolutely irresistible, if you make for the enemy's ___(ii)___ points; you may retire and be ___(iii)___ from pursuit if your movements are more ___(iv)___ than those of the enemy.

(i)	(ii)	(iii)	(iv)
(a) weak	advance	rapid	safe
(b) weak	advance	safe	rapid
(c) advance	weak	safe	rapid
(d) advance	weak	rapid	safe

Q9. Sun Tzu said: If we wish to ___(i)___, the enemy can be ___(ii)___ to an engagement even though he be sheltered behind a high rampart and a deep ditch. All we need do is ___(iii)___ some other place that he will be obliged to ___(iv)___.

(i)	(ii)	(iii)	(iv)
(a) relieve	forced	fight	attack
(b) fight	forced	attack	relieve
(c) fight	relieve	attack	forced
(d) attack	forced	fight	relieve

Q10. Sun Tzu said: By discovering the enemy's ___(i)___ and remaining ___(ii)___ ourselves, we can keep our forces ___(iii)___, while the enemy's must be ___(iv)___.

(i)	(ii)	(iii)	(iv)
(a) invisible	concentrated	dispositions	divided
(b) invisible	concentrated	divided	dispositions
(c) dispositions	invisible	concentrated	divided
(d) dispositions	invisible	divided	concentrated

Q11. Sun Tzu said: We can form a single ___(i)___ body, while the enemy must ___(ii)___ up into fractions. Hence there will be a ___(iii)___ pitted against separate ___(iv)___ of a whole, which means that we shall be many to the enemy's few.

(i)	(ii)	(iii)	(iv)
(a) whole	parts	united	split
(b) parts	whole	united	split
(c) united	split	parts	whole
(d) united	split	whole	parts

Q12. Sun Tzu said: The spot where we intend to fight must not be made ___(i)___; for then the enemy will have to prepare against a possible attack at ___(ii)___ different points; and his forces being thus ___(iii)___ in many directions, the numbers we shall have to face at any given point will be proportionately ___(iv)___.

(i)	(ii)	(iii)	(iv)
(a) several	known	few	distributed
(b) several	known	distributed	few
(c) known	several	few	distributed
(d) known	several	distributed	few

Q13. Sun Tzu said: If enemy sends reinforcements everywhere, he will everywhere be strong. (True or False)

Q14. Sun Tzu said: Numerical ___(i)___ comes from having to prepare against possible attacks; numerical ___(ii)___, from compelling our adversary to make these preparations against us.

(i)	(ii)
(a) strength	weakness
(b) weakness	weakness
(c) strength	strength
(d) weakness	strength

Q15. Sun Tzu said: Knowing the place and the time of the coming battle, we may concentrate from the greatest distances in order to fight.(True or False)

Q16 Sun Tzu said: If neither time nor place be known, then -

(a) the left wing will be impotent to succor the right
(b) the right wing will be impotent to succor the left
(c) the van will be unable to relieve the rear
(d) the rear will be unable to support the van

Q17. Though the enemy be stronger in numbers, we may prevent him from fighting. How?

(a) Scheme so as to discover his plans and the likelihood of their success.
(b) Rouse him, and learn the principle of his activity or inactivity.
(c) Force him to reveal himself, so as to find out his vulnerable spots.
(d) Carefully compare the opposing army with your own, so that you may know where strength is superabundant and where it is deficient.

Q18. Sun Tzu said: In making tactical dispositions, the highest pitch you can attain is to ___(i)___ them; ___(ii)___ your dispositions, and you will be ___(iii)___ from the prying of the subtlest ___(iv)___, from the machinations of the wisest brains.

(i)	(ii)	(iii)	(iv)
(a) conceal	spies	conceal	safe
(b) conceal	spies	safe	conceal
(c) spies	conceal	conceal	safe
(d) conceal	conceal	safe	spies

Q19. Sun Tzu said: How ___(i)___ may be produced for them out of the ___(ii)___ own ___(iii)___ - that is what the ___(iv)___ cannot comprehend.

(i)	(ii)	(iii)	(iv)
(a) tactics	multitude	victory	enemy's
(b) tactics	multitude	enemy's	victory
(c) victory	enemy's	tactics	multitude
(d) victory	multitude	tactics	enemy's

Q20. Sun Tzu said: All men can see the ___(i)___ whereby I ___(ii)___, but what none can see is the ___(iii)___ out of which ___(iv)___ is evolved.

	(i)	(ii)	(iii)	(iv)
(a)	strategy	victory	tactics	conquer
(b)	tactics	conquer	strategy	victory
(c)	strategy	conquer	tactics	victory
(d)	victory	conquer	tactics	strategy

Q21. Sun Tzu said: Do not repeat the tactics which have gained you one victory, but let your methods be regulated by the infinite variety of circumstances. (True or False)

Q22. Sun Tzu said: Military ___(i)___ are like unto ___(ii)___; for water in its natural course runs away from high places and hastens downwards. So in war, the way is to avoid what is ___(iii)___ and to strike at what is ___(iv)___.

	(i)	(ii)	(iii)	(iv)
(a)	water	tactics	weak	strong
(b)	weak	water	tactics	strong
(c)	tactics	water	strong	weak
(d)	tactics	water	weak	strong

Q23. Sun Tzu said: Water shapes its ___(i)___ according to the ___(ii)___ of the ground over which it flows; the soldier works out his ___(iii)___ in relation to the ___(iv)___ whom he is facing.

	(i)	(ii)	(iii)	(iv)
(a)	nature	course	foe	victory
(b)	foe	nature	course	victory
(c)	victory	course	foe	nature
(d)	course	nature	victory	foe

Q24. Sun Tzu said: Just as water retains no constant ___(i)___, so in warfare there are no constant ___(ii)___. He who can ___(iii)___ his tactics in relation to his ___(iv)___ and thereby succeed in winning, may be called a heaven-born captain.

	(i)	(ii)	(iii)	(iv)
(a)	conditions	modify	shape	opponent
(b)	conditions	opponent	shape	modify
(c)	shape	conditions	modify	opponent
(d)	shape	opponent	modify	conditions

6

WEAK POINTS AND STRONG

Answers

Q1. (d)
Q2. True
Q3. (a), (b), (c), (d)
Q4. (d)
Q5. True
Q6. (c)
Q7. (d)
Q8. (c)
Q9. (b)
Q10. (c)
Q11. (d)
Q12. (d)
Q13. False
Q14. (d)
Q15. True
Q16. (a), (b), (c), (d)
Q17. (a), (b), (c), (d)
Q18. (d)
Q19. (c)
Q20. (b)
Q21. True
Q22. (c)
Q23. (d)
Q24. (c)

7

MANEUVERING

Instructions:
1. Select the appropriate answer(s) from the questions with four options (a), (b), (c) and (d). In some questions there can be more than one correct answer.
2. In the questions with True or False options, choose the appropriate option.

Q1. Sun Tzu said: In war, the general having collected an army and concentrated his forces, he must ___(i)___ the different elements thereof before pitching his ___(ii)___. After that, comes ___(iii)___, than which there is nothing more difficult.

(i)	(ii)	(iii)
(a) camp	tactical maneuvering	blend and harmonize
(b) blend and harmonize	tactical maneuvering	camp
(c) camp	blend and harmonize	tactical maneuvering
(d) blend and harmonize	camp	tactical maneuvering

Q2. Sun Tzu said: The ___(i)___ of tactical maneuvering consists in turning the devious into the ___(ii)___, and misfortune into ___(iii)___.

(i)	(ii)	(iii)
(a) gain	direct	difficulty
(b) difficulty	direct	gain
(c) gain	direct	difficulty
(d) difficulty	gain	direct

Q3. Sun Tzu said: To take a long and circuitous route, after enticing the ___(i)___ out of the way, and though starting after him, to contrive to reach the ___(ii)___ before him, shows ___(iii)___ of the artifice of ___(iv)___ .

	(i)	(ii)	(iii)	(iv)
(a)	goal	enemy	deviation	knowledge
(b)	goal	knowledge	deviation	enemy
(c)	enemy	goal	knowledge	deviation
(d)	enemy	knowledge	deviation	goal

Q4. Sun Tzu said: ___(i)___ with an army is ___(ii)___; with an ___(iii)___ multitude, most ___(iv)___ .

	(i)	(ii)	(iii)	(iv)
(a)	undisciplined	dangerous	maneuvering	advantageous
(b)	undisciplined	advantageous	maneuvering	dangerous
(c)	maneuvering	dangerous	undisciplined	advantageous
(d)	maneuvering	advantageous	undisciplined	dangerous

Q5. Sun Tzu said: If you set a fully equipped ___(i)___ in march in order to snatch an ___(ii)___, the chances are that you will be too late. On the other hand, to detach a ___(iii)___ for the purpose involves the ___(iv)___ of its baggage and stores.

	(i)	(ii)	(iii)	(iv)
(a)	flying column	army	sacrifice	advantage
(b)	flying column	army	advantage	sacrifice
(c)	army	advantage	flying column	sacrifice
(d)	army	sacrifice	flying column	advantage

Q6. Sun Tzu said: We cannot enter into ___(i)___ until we are ___(ii)___ with the ___(iii)___ of our ___(iv)___ .

	(i)	(ii)	(iii)	(iv)
(a)	designs	acquainted	neighbors	alliances
(b)	neighbors	alliances	designs	acquainted

(c) designs   alliances   neighbors   acquainted
(d) alliances   acquainted   designs   neighbors

Q7. Sun Tzu said: We are not fit to ___(i)___ an army on the ___(ii)___ unless we are familiar with the ___(iii)___ of the ___(iv)___ - its mountains and forests, its pitfalls and precipices, its marshes and swamps.

(i)	(ii)	(iii)	(iv)
(a) march	lead	country	face
(b) lead	march	face	country
(c) march	lead	face	country
(d) lead	country	march	face

Q8. Sun Tzu said: We shall be unable to turn natural advantage to account unless we make use of local guides. (True or False)

Q9. Sun Tzu said: In war, practice dissimulation, and you will not succeed. (True or False)

Q10. Sun Tzu said: Whether to concentrate or to divide your troops, must be decided by circumstances. (True or False)

Q11. Sun Tzu did not say:

(a) In raiding and plundering be like wind, in immovability like a mountain.
(b) Let your plans be dark and impenetrable as night.
(c) Let your rapidity be that of the wind, your compactness that of the forest.
(d) When you move, fall like a thunderbolt.

Q12. Sun Tzu said: When you plunder a countryside, let the spoil be ___(i)___ amongst your ___(ii)___; when you capture new territory, cut it up into allotments for the ___(iii)___ of the ___(iv)___.

(i)	(ii)	(iii)	(iv)
(a) benefit	men	soldiery	divided
(b) men	soldiery	benefit	divided

| (c) divided | men | benefit | soldiery |
| (d) divided | benefit | men | soldiery |

Q13. Sun Tzu said: Ponder and deliberate before you make a move. (True or False)

Q14. Sun Tzu said: He will ___(i)___ who has ___(ii)___ the artifice of ___(iii)___. Such is the art of ___(iv)___.

(i)	(ii)	(iii)	(iv)
(a) learnt	conquer	maneuvering	deviation
(b) learnt	conquer	deviation	maneuvering
(c) conquer	learnt	maneuvering	deviation
(d) conquer	learnt	deviation	maneuvering

Q15. Sun Tzu said: The Book of Army Management says: On the field of battle, the ___(i)___ does not carry far enough: hence the institution of ___(ii)___. Nor can ordinary objects be ___(iii)___ clearly enough: hence the institution of ___(iv)___.

(i)	(ii)	(iii)	(iv)
(a) gongs and drums	spoken word	seen	banners and flags
(b) banners and flags	spoken word	seen	gongs and drums
(c) spoken word	gongs and drums	seen	banners and flags
(d) spoken word	banners and flags	gongs and drums	seen

Q16. Sun Tzu said: The host (army) forming a ___(i)___ united body, it is impossible either for the brave to ___(ii)___ alone, or for the cowardly to ___(iii)___ alone. This is the art of handling ___(iv)___ masses of men.

(i)	(ii)	(iii)	(iv)
(a) large	retreat	advance	single
(b) single	advance	retreat	large
(c) single	retreat	advance	large
(d) large	advance	retreat	single

Q17. Sun Tzu said: In night-fighting make much use of ___(i)___, and in fighting by day, of ___(ii)___, as a means of influencing the ___(iii)___ of your army.

(i)	(ii)	(iii)
(a) flags and banners	ears and eyes	signal-fires and drums
(b) flags and banners	signal-fires and drums	ears and eyes
(c) signal-fires and drums	ears and eyes	flags and banners
(d) signal-fires and drums	flags and banners	ears and eyes

Q18. Sun Tzu said: A whole army may be robbed of its spirit; a commander-in-chief may be robbed of his presence of mind. (True or False)

Q19. Sun Tzu said: A soldier's spirit is ___(i)___ in the morning; by noonday it has begun to ___(ii)___; and in the evening, his mind is bent only on ___(iii)___ to camp.

(i)	(ii)	(iii)
(a) returning	flag	keenest
(b) returning	keenest	flag
(c) keenest	flag	returning
(d) flag	keenest	returning

Q20. Sun Tzu said: A clever general ___(i)___ an army when its spirit is ___(ii)___, but ___(iii)___ it when it is ___(iv)___ and inclined to return to camp. This is the art of studying moods.

(i)	(ii)	(iii)	(iv)
(a) attacks	keen	avoids	sluggish
(b) attacks	sluggish	avoids	keen
(c) avoids	sluggish	attacks	keen
(d) avoids	keen	attacks	sluggish

Q21. Sun Tzu said: Disciplined and calm, to ___(i)___ the appearance of ___(ii)___ and hubbub amongst the ___(iii)___:- this is the art of retaining ___(iv)___.

(i)	(ii)	(iii)	(iv)
(a) disorder	self-possession	await	enemy
(b) await	disorder	enemy	self-possession
(c) await	self-possession	disorder	enemy
(d) disorder	self-possession	enemy	await

Q22. Sun Tzu said: To be ___(i)___ the goal while the enemy is still ___(ii)___ from it, to wait at ___(iii)___ while the enemy is toiling and ___(iv)___, to be well-fed while the enemy is famished :- this is the art of husbanding one's strength.

(i)	(ii)	(iii)	(iv)
(a) struggling	far	ease	near
(b) near	struggling	far	ease
(c) struggling	far	ease	near
(d) near	far	ease	struggling

Q23. Which of the following is (are) true of the art of studying circumstances?

(a) Refrain from intercepting an enemy whose banners are in perfect order.
(b) Attack an army drawn up in calm and confident array.
(c) Intercept an enemy whose banners are in perfect order.
(d) Refrain from attacking an army drawn up in calm and confident array.

Q24. Sun Tzu said: It is a military axiom not to ___(i)___ against the enemy, nor to ___(ii)___ him when he comes ___(iii)___.

(i)	(ii)	(iii)
(a) oppose	downhill	advance uphill
(b) advance uphill	oppose	downhill
(c) advance uphill	downhill	advance uphill
(d) downhill	oppose	advance uphill

Q25. Sun Tzu did not say:

(a) Do not pursue an enemy who simulates flight.
(b) Do attack soldiers whose temper is keen.
(c) Do not swallow bait offered by the enemy.
(d) Do not interfere with an army that is returning home.

Q26. Sun Tzu said: When you ___(i)___ an army, ___(ii)___ an outlet free. Do not ___(iii)___ a desperate foe too hard. Such is the art of warfare.

(i)	(ii)	(iii)
(a) leave	press	surround
(b) press	surround	leave
(c) surround	press	leave
(d) surround	leave	press

7

MANEUVERING

Answers

Q1. (d)
Q2. (b)
Q3. (c)
Q4. (d)
Q5. (c)
Q6. (d)
Q7. (b)
Q8. True
Q9. False
Q10. True
Q11. (a)
Q12. (c)
Q13. True
Q14. (d)
Q15. (c)
Q16. (b)
Q17. (d)
Q18. True
Q19. (c)
Q20. (d)
Q21. (b)
Q22. (d)
Q23. (a), (d)
Q24. (b)
Q25. (b)
Q26. (d)

8
VARIATION IN TACTICS

Instructions:
1. Select the appropriate answer(s) from the questions with four options (a), (b), (c) and (d). In some questions there can be more than one correct answer.
2. In the questions with True or False options, choose the appropriate option.

Q1. Sun Tzu said: In war, the general receives his commands from the sovereign, collects his army and concentrates his forces. (True or False)

Q2. Sun Tzu said: When in difficult country, do not ___(i)___. In country where high roads intersect, join hands with your ___(ii)___. Do not linger in dangerously ___(iii)___ positions. In hemmed-in situations, you must resort to stratagem. In desperate position, you must ___(iv)___.

(i)	(ii)	(iii)	(iv)
(a) fight	isolated	allies	encamp
(b) fight	isolated	encamp	allies
(c) encamp	allies	isolated	fight
(d) encamp	isolated	allies	fight

Q3. Sun Tzu said: There are roads which must not be ___(i)___, armies which must be not ___(ii)___, towns which must be ___(iii)___, positions which must not be ___(iv)___, commands of the sovereign which must not be obeyed.

58

(i)	(ii)	(iii)	(iv)
(a) besieged	contested	attacked	followed
(b) followed	contest	attacked	besieged
(c) besieged	contested	followed	attacked
(d) followed	attacked	besieged	contested

Q4. Sun Tzu said: The general who thoroughly understands the ___(i)___ that accompany ___(ii)___ of tactics knows how to handle his troops. The general who does not understand these, may be well acquainted with the configuration of the ___(iii)___, yet he will not be able to turn his knowledge to ___(iv)___ account.

(i)	(ii)	(iii)	(iv)
(a) country	advantages	variation	practical
(b) variation	practical	advantages	country
(c) advantages	variation	country	practical
(d) advantages	practical	variation	country

Q5. Sun Tzu said: The student of war who is well versed in the art of war of varying his plans will fail to make the best use of his men. (True or False)

Q6. Sun Tzu said: In the wise leader's plans, considerations of advantage and of disadvantage will be blended together. (True or False)

Q7. Sun Tzu said: ___(i)___ the hostile chiefs by inflicting ___(ii)___ on them; and make ___(iii)___ for them, and keep them constantly ___(iv)___; hold out specious allurements, and make them rush to any given point.

(i)	(ii)	(iii)	(iv)
(a) trouble	reduce	damage	engaged
(b) reduce	damage	trouble	engaged
(c) reduce	trouble	damage	engaged
(d) engaged	trouble	reduce	damage

Q8. Sun Tzu said: The art of war teaches us to rely not on the likelihood of the enemy's ___(i)___, but on our own ___(ii)___ to receive him; not on the chance of his ___(iii)___ , but rather on the fact that we have made our position ___(iv)___ .

(i)	(ii)	(iii)	(iv)
(a) readiness	unassailable	not coming	not attacking
(b) unassailable	readiness	not attacking	not coming
(c) not coming	unassailable	not attacking	readiness
(d) not coming	readiness	not attacking	unassailable

Q9. Which of the following are besetting sin of a general, ruinous to the conduct of war ?

(a) Recklessness, which leads to destruction
(b) Cowardice, which leads to capture
(c) A hasty temper, which can be provoked by insults
(d) Over-solicitude for his men, which exposes him to worry and trouble.

Q10. Sun Tzu said: When an army is overthrown and its leader slain, the cause will surely be found among which dangerous faults?

(a) Recklessness
(b) Cowardice
(c) A hasty temper
(d) Over-solicitude for his men

8

VARIATION IN TACTICS

Answers

Q1. True

Q2. (c)

Q3. (d)

Q4. (c)

Q5. False

Q6. True

Q7. (b)

Q8. (d)

Q9. (a), (b), (c), (d)

Q10. (a), (b), (c), (d)

9

THE ARMY ON THE MARCH

Instructions:
1. Select the appropriate answer(s) from the questions with four options (a), (b), (c) and (d). In some questions there can be more than one correct answer.
2. In the questions with True or False options, choose the appropriate option.

Q1. Sun Tzu said: ___(i)___ in high places, facing the sun. Do not climb ___(ii)___ in order to ___(iii)___. So much for ___(iv)___ warfare.

(i)	(ii)	(iii)	(iv)
(a) fight	mountain	camp	heights
(b) fight	heights	camp	mountain
(c) camp	mountain	fight	heights
(d) camp	heights	fight	mountain

Q2. Which of the following is (are) true of river warfare?

(a) When an invading force crosses a river in its onward march, do not advance to meet it in mid-stream. It will be best to let half the army get across, and then deliver your attack.
(b) After crossing a river, you should get far away from it.
(c) If you are anxious to fight, you should not go to meet the invader near a river which he has to cross.
(d) Moor your craft higher up than the enemy, and facing the sun. Do not move down-stream to meet the enemy.

Q3. Which of the following is (are) true of operations in salt-marshes?

(a) In crossing salt-marshes, your sole concern should be to get over them quickly, without any delay.
(b) Encamp at salt-marshes without any delay.
(c) If forced to fight in a salt-marsh, you should have water and grass near you, and get your back to a clump of trees.
(d) If forced to fight in a salt-marsh, you should get your back to a clump of trees.

Q4. Sun Tzu said: In campaigning in dry, level country, take up an easily accessible position with rising ground to your right and on your ___(i)___, so that the ___(ii)___ may be in ___(iii)___, and ___(iv)___ lie behind.

(i)	(ii)	(iii)	(iv)
(a) front	safety	rear	danger
(b) front	danger	rear	safety
(c) rear	danger	front	safety
(d) rear	safety	front	danger

Q5. What should be done in a country in which there are precipitous cliffs with torrents running between, deep natural hollows, confined places, tangled thickets, quagmires and crevasses?

(a) Keep away from such places.
(b) Get the enemy to approach such places.
(c) Leave such places with all possible speed.
(d) While such places are in our front, we should let the enemy have them on his rear.

Q6. Sun Tzu said: If in the ___(i)___ of your camp there should be any hilly country, ponds surrounded by aquatic grass, hollow basins filled with reeds, or woods with thick undergrowth, they must be carefully routed out and ___(ii)___; for these are places where men in ___(iii)___ or insidious ___(iv)___ are likely to be lurking.

	(i)	(ii)	(iii)	(iv)
(a)	ambush	searched	neighborhood	spies
(b)	spies	ambush	neighborhood	searched
(c)	neighborhood	searched	ambush	spies
(d)	ambush	neighborhood	searched	spies

Q7. Sun Tzu said: When the enemy is ___(i)___ at hand and remains ___(ii)___, he is relying on the natural ___(iii)___ of his ___(iv)___ .

	(i)	(ii)	(iii)	(iv)
(a)	quiet	close	position	strength
(b)	close	quiet	position	strength
(c)	quiet	close	strength	position
(d)	close	quiet	strength	position

Q8. Sun Tzu said: When enemy keeps ___(i)___ and tries to ___(ii)___ a battle, he is anxious for the ___(iii)___ to ___(iv)___ .

	(i)	(ii)	(iii)	(iv)
(a)	other side	advance	aloof	provoke
(b)	aloof	provoke	other side	advance
(c)	other side	provoke	aloof	advance
(d)	aloof	advance	other side	provoke

Q9. Sun Tzu said: If place of encampment of enemy is easy of access, he is tendering a bait. (True or False)

Q10. Sun Tzu said: ___(i)___ amongst the trees of a forest shows that the enemy is ___(ii)___ . The appearance of a number of screens in the midst of thick grass means that the ___(iii)___ wants to make us ___(iv)___ .

	(i)	(ii)	(iii)	(iv)
(a)	enemy	advancing	suspicious	movement
(b)	suspicious	enemy	advancing	movement
(c)	movement	advancing	enemy	suspicious
(d)	advancing	movement	suspicious	enemy

Q11. Sun Tzu said: The ___(i)___ of birds in their flight is the sign of an ___(ii)___. ___(iii)___ beasts indicate that a sudden ___(iv)___ is coming.

(i)	(ii)	(iii)	(iv)
(a) attack	ambuscade	startled	rising
(b) ambuscade	startled	rising	attack
(c) attack	ambuscade	rising	startled
(d) rising	ambuscade	startled	attack

Q12. Sun Tzu said: When there is dust ___(i)___ in a high column, it is the sign of ___(ii)___ advancing; when the dust is ___(iii)___, but spread over a wide area, it betokens the approach of ___(iv)___.

(i)	(ii)	(iii)	(iv)
(a) low	infantry	rising	chariots
(b) rising	chariots	low	infantry
(c) low	chariots	rising	infantry
(d) rising	infantry	low	chariots

Q13. Sun Tzu said: When dust branches out in ___(i)___ directions, it shows that parties have been ___(ii)___ to collect firewood. A few clouds of dust moving ___(iii)___ signify that the army is ___(iv)___.

(i)	(ii)	(iii)	(iv)
(a) to and fro	encamping	different	sent
(b) different	encamping	to and fro	sent
(c) to and fro	sent	different	encamping
(d) different	sent	to and fro	encamping

Q14. Sun Tzu said: ___(i)___ words and increased preparations are signs that the enemy is about to ___(ii)___. ___(iii)___ language and driving forward as if to the attack are signs that he will ___(iv)___.

(i)	(ii)	(iii)	(iv)
(a) violent	advance	humble	retreat
(b) advance	retreat	violent	humble

| (c) violent | retreat | humble | advance |
| (d) humble | advance | violent | retreat |

Q15. Sun Tzu said: When the light chariots come out first and take up a position on the wings, it is a sign that the enemy is forming for battle. (True or False)

Q16. Sun Tzu said: ___(i)___ proposals ___(ii)___ by a ___(iii)___ covenant indicate a ___(iv)___.

(i)	(ii)	(iii)	(iv)
(a) unaccompanied	sworn	peace	plot
(b) sworn	plot	unaccompanied	peace
(c) peace	unaccompanied	sworn	plot
(d) unaccompanied	plot	sworn	peace

Q17. Sun Tzu said: When there is much running about and the soldiers fall out of rank, it means that the critical moment has come. (True or False)

Q18. Sun Tzu said: When some soldiers are seen advancing and some retreating, it is a lure. (True or False)

Q19. Sun Tzu said: If the soldiers who are sent to draw water begin by drinking themselves, the army is suffering from thirst. (True or False)

Q20. Sun Tzu said: If the enemy sees an advantage to be gained and makes no effort to secure it, the soldiers are exhausted. (True or False)

Q21. Sun Tzu said: If birds gather on any spot, it is unoccupied. (True or False)

Q22. Sun Tzu said: If there is disturbance in the camp, the general's authority is ___(i)___. If the banners and flags are shifted about, ___(ii)___ is afoot. If the officers are ___(iii)___, it means that the men are ___(iv)___.

	(i)	(ii)	(iii)	(iv)
(a)	weary	sedition	weak	angry
(b)	angry	weak	sedition	weary
(c)	weak	sedition	angry	weary
(d)	sedition	weary	weak	angry

Q23. Sun Tzu said: The sight of men whispering together in small knots or speaking in subdued tones points to disaffection amongst the rank and file. (True or False)

Q24. Sun Tzu said: To begin by ___(i)___, but afterwards to take ___(ii)___ at the enemy's ___(iii)___, shows a supreme lack of ___(iv)___.

	(i)	(ii)	(iii)	(iv)
(a)	fright	bluster	intelligence	numbers
(b)	bluster	numbers	intelligence	fright
(c)	intelligence	fright	bluster	numbers
(d)	bluster	fright	numbers	intelligence

Q25. Sun Tzu said: When ___(i)___ are sent with ___(ii)___ in their mouths, it is a sign that the ___(iii)___ wishes for a ___(iv)___.

	(i)	(ii)	(iii)	(iv)
(a)	truce	enemy	envoys	compliments
(b)	enemy	truce	envoys	compliments
(c)	envoys	compliments	enemy	truce
(d)	compliments	truce	envoys	enemy

Q26. If our troops are no more in number than the enemy, that is amply sufficient, what can we do?

(a) Make no direct attack on enemy
(b) Concentrate all our available strength
(c) Keep a close watch on the enemy
(d) Obtain reinforcements

Q27. Sun Tzu said: He who exercises no ___(i)___ but makes ___(ii)___ of his ___(iii)___ is sure to be ___(iv)___ by them.

	(i)	(ii)	(iii)	(iv)
(a)	opponents	forethought	captured	light
(b)	forethought	light	opponents	captured
(c)	opponents	light	captured	forethought
(d)	light	opponents	captured	forethought

Q28. Sun Tzu said: If soldiers are ___(i)___ before they have grown ___(ii)___ to you, they will not prove ___(iii)___; and, unless submissive, then will be practically ___(iv)___.

	(i)	(ii)	(iii)	(iv)
(a)	useless	attached	submissive	punished
(b)	attached	submissive	useless	punished
(c)	submissive	useless	attached	punished
(d)	punished	attached	submissive	useless

Q29. Sun Tzu said: If, when the soldiers have become ___(i)___ to you, ___(ii)___ are not ___(iii)___, they will still be ___(iv)___.

	(i)	(ii)	(iii)	(iv)
(a)	useless	attached	punishments	enforced
(b)	punishments	useless	attached	enforced
(c)	attached	punishments	enforced	useless
(d)	enforced	punishments	useless	attached

Q30. Sun Tzu said: Soldiers must be treated in the first instance with ___(i)___, but kept under ___(ii)___ by means of iron ___(iii)___. This is a certain road to ___(iv)___.

	(i)	(ii)	(iii)	(iv)
(a)	victory	discipline	control	humanity
(b)	discipline	control	victory	humanity
(c)	humanity	control	discipline	victory

(d) humanity   victory   control   discipline

Q31. Sun Tzu said: If in ___(i)___ soldiers ___(ii)___ are habitually ___(iii)___, the army will be well-disciplined; if not, its ___(iv)___ will be bad.

(i)	(ii)	(iii)	(iv)
(a) enforced	training	commands	discipline
(b) discipline	training	enforced	commands
(c) enforced	commands	discipline	training
(d) training	commands	enforced	discipline

9

THE ARMY ON THE MARCH

Answers

Q1. (d)

Q2. (a), (b), (c)

Q3. (a), (c), (d)

Q4. (c)

Q5. (a), (b), (c), (d)

Q6. (c)

Q7. (d)

Q8. (b)

Q9. True

Q10. (c)

Q11. (d)

Q12. (b)

Q13. (d)

Q14. (d)

Q15. True

Q16. (c)

Q17. False

Q18. True

Q19. True

Q20. True

Q21. True

Q22. (c)

Q23. True

Q24. (d)

Q25. (c)

Q26. (a), (b), (c), (d)

Q27. (b)

Q28. (d)

Q29. (c)

Q30. (c)

Q31. (d)

10
TERRAIN

Instructions:
1. Select the appropriate answer(s) from the questions with four options (a), (b), (c) and (d). In some questions there can be more than one correct answer.
2. In the questions with True or False options, choose the appropriate option.

Q1. Sun Tzu said: We may distinguish six kinds of terrain, to wit: (1) Accessible ground; (2) ___(i)___; (3) temporizing ground; (4) ___(ii)___; (5) ___(iii)___; (6) positions at a great distance from the enemy.

(i)	(ii)	(iii)
(a) narrow passes	entangling ground	desperate ground
(b) entangling ground	narrow passes	precipitous heights
(c) precipitous heights	serious ground	narrow passes
(d) desperate ground	serious ground	entangling ground

Q2. Which of the following is (are) true of Accessible ground ?

(a) Be before the enemy in occupying the raised and sunny spots on accessible ground.
(b) Accessible ground can be freely traversed by both sides.
(c) With regard to ground of this nature, carefully guard your line of supplies.
(d) None of these.

Q3. Which of the following is (are) true of Entangling ground?

(a) Entangling ground can be abandoned but is hard to re-occupy.

(b) If the enemy on entangling is unprepared, you may sally forth and defeat him.

(c) If the enemy on entangling is prepared for your coming, and you fail to defeat him, then, return being impossible, disaster will ensue.

(d) Only (a) and (c)

Q4. Sun Tzu said: When the position is such that neither side will gain by making _____ move, it is called temporizing ground.

(a) any

(b) no

(c) the first

(d) the second

Q5. Sun Tzu said: In a position of temporizing ground, even though the enemy should offer us an attractive ___(i)___, it will be advisable not to stir forth, but rather to ___(ii)___, thus ___(iii)___ the enemy in his turn; then, when part of his army has come out, we may deliver our ___(iv)___ with advantage.

	(i)	(ii)	(iii)	(iv)
(a)	enticing	attack	retreat	bait
(b)	retreat	bait	enticing	attack
(c)	attack	retreat	bait	enticing
(d)	bait	retreat	enticing	attack

Q6. Sun Tzu said: With regard to narrow passes, if you can ___(i)___ them ___(ii)___, let them be strongly garrisoned and ___(iii)___ the advent of the ___(iv)___.

	(i)	(ii)	(iii)	(iv)
(a)	await	first	occupy	enemy
(b)	await	enemy	occupy	first
(c)	occupy	first	await	enemy
(d)	occupy	enemy	await	first

Q7. Sun Tzu said: Should the army forestall you in occupying a pass, do not go after him if the pass is ___(i)___ garrisoned, but only if it is ___(ii)___ garrisoned.

(i)	(ii)
(a) fully	fully
(b) weakly	fully
(c) fully	weakly
(d) weakly	weakly

Q8. Which of the following is (are) true of Precipitous heights ?

(a) If you reach precipitous heights before your adversary, you should occupy the raised and sunny spots, and there wait for him to come up.
(b) If the enemy has occupied precipitous heights before you, follow him and don't retreat.
(c) If the enemy has occupied them before you, do not follow him, but retreat and try to entice him away.
(d) None of these.

Q9. Sun Tzu said: If you are situated at a great ___(i)___ from the enemy, and the strength of the two armies is ___(ii)___, it is not easy to provoke a battle, and ___(iii)___ will be to your ___(iv)___.

(i)	(ii)	(iii)	(iv)
(a) disadvantage	fighting	equal	distance
(b) distance	fighting	equal	disadvantage
(c) equal	distance	disadvantage	fighting
(d) distance	equal	fighting	disadvantage

Q10. Sun Tzu said: There are six principles connected with Earth. The general who has attained a responsible post must be careful to study them. (True or False)

Q11. Sun Tzu said: An army is exposed to six several calamities, not arising from natural causes, but from faults for which the general is responsible.

These are: (1) Flight; (2) ___(i)___; (3) collapse; (4) ___(ii)___; (5) disorganization; (6) ___(iii)___.

(i)	(ii)	(iii)
(a) ruin	attack	rout
(b) insubordination	ruin	retreat
(c) rout	attack	insubordination
(d) insubordination	ruin	rout

Q12. Sun Tzu said: Other conditions being equal, if one force is hurled against another ten times its size, the result will be the flight of the former. (True or False)

Q13. Sun Tzu said: When the common soldiers are too ___(i)___ and their officers too ___(ii)___, the result is insubordination. When the officers are too ___(iii)___ and the common soldiers too ___(iv)___, the result is collapse.

(i)	(ii)	(iii)	(iv)
(a) weak	strong	strong	weak
(b) strong	weak	strong	weak
(c) strong	weak	weak	strong
(d) weak	strong	weak	strong

Q14. Sun Tzu said: When the higher officers are ___(i)___ and insubordinate, and on meeting the ___(ii)___ give battle on their own account from a feeling of ___(iii)___, before the commander-in-chief can tell whether or no he is in a position to ___(iv)___, the result is ruin.

(i)	(ii)	(iii)	(iv)
(a) enemy	angry	fight	resentment
(b) resentment	angry	enemy	fight
(c) angry	enemy	resentment	fight
(d) angry	fight	resentment	enemy

Q15. The result of which of the following is (are) utter disorganization?

(a) When the general is weak and without authority.
(b) When the general's orders are not clear and distinct.
(c) When there are no fixed duties assigned to officers and men.
(d) When the ranks are formed in a slovenly haphazard manner.

Q16. Sun Tzu said: When a general, unable to estimate the enemy's strength, allows an ___(i)___ force to engage a ___(ii)___ one, or hurls a ___(iii)___ detachment against a ___(iv)___ one, and neglects to place picked soldiers in the front rank, the result must be rout.

(i)	(ii)	(iii)	(iv)
(a) larger	inferior	powerful	weak
(b) powerful	weak	larger	inferior
(c) weak	inferior	powerful	larger
(d) inferior	larger	weak	powerful

Q17. Sun Tzu said: The natural formation of the country is the soldier's best ally; but a power of ___(i)___ the adversary, of ___(ii)___ the forces of victory, and of shrewdly calculating ___(iii)___, dangers and distances, constitutes the ___(iv)___ of a great general. He who knows these things, and in fighting puts his knowledge into practice, will win his battles. He who knows them not, nor practices them, will surely be defeated.

(i)	(ii)	(iii)	(iv)
(a) controlling	estimating	test	difficulties
(b) estimating	controlling	difficulties	test
(c) test	estimating	controlling	difficulties
(d) controlling	estimating	difficulties	test

Q18. Sun Tzu said: If fighting is sure to result in victory, then you ___(i)___ fight, even though the ruler ___(ii)___ it; if fighting will not result in victory, then you ___(iii)___ fight even at the ruler's ___(iv)___.

(i)	(ii)	(iii)	(iv)
(a) must not	forbid	must	bidding
(b) must not	bidding	must	forbid
(c) must	forbid	must not	bidding
(d) forbid	must	must not	bidding

Q19. Sun Tzu said: The general who ___(i)___ without coveting ___(ii)___ and ___(iii)___ without fearing ___(iv)___, whose only thought is to protect his country and do good service for his sovereign, is the jewel of the kingdom.

(i)	(ii)	(iii)	(iv)
(a) retreats	fame	advances	disgrace
(b) fame	disgrace	advances	retreats
(c) retreats	disgrace	advances	fame
(d) advances	fame	retreats	disgrace

Q20. Sun Tzu said: Regard your soldiers as your children, and they will follow you into the deepest valleys; look upon them as your own beloved sons, and they will stand by you even unto death. (True or False)

Q21. Sun Tzu said: If you are indulgent, but unable to make your ___(i)___ felt; kind-hearted, but unable to enforce your ___(ii)___; and incapable, moreover, of quelling ___(iii)___: then your soldiers must be likened to spoilt children; they are ___(iv)___ for any practical purpose.

(i)	(ii)	(iii)	(iv)
(a) disorder	authority	useless	commands
(b) commands	authority	useless	disorder
(c) useless	commands	disorder	authority
(d) authority	commands	disorder	useless

Q22. Sun Tzu said: If we ___(i)___ that our own men are in a ___(ii)___ to attack, but are ___(iii)___ that the enemy is ___(iv)___ to attack, we have gone only halfway towards victory.

	(i)	(ii)	(iii)	(iv)
(a)	condition	not open	unaware	know
(b)	unaware	not open	condition	know
(c)	condition	unaware	know	not open
(d)	know	condition	unaware	not open

Q23. Sun Tzu said: If we ___(i)___ that the enemy is ___(ii)___ to attack, but are ___(iii)___ that our own men are ___(iv)___ to attack, we have gone only halfway towards victory.

	(i)	(ii)	(iii)	(iv)
(a)	open	know	not in a condition	unaware
(b)	know	open	unaware	not in a condition
(c)	know	open	not in a condition	unaware
(d)	open	know	unaware	not in a condition

Q24. Sun Tzu said: If we ___(i)___ that the enemy is ___(ii)___ to attack, and also know that our men are in a condition to attack, but are ___(iii)___ that the nature of the ___(iv)___ makes fighting impracticable, we have gone only halfway towards victory.

	(i)	(ii)	(iii)	(iv)
(a)	open	ground	know	unaware
(b)	unaware	ground	open	know
(c)	know	open	unaware	ground
(d)	open	unaware	know	ground

Q25. Sun Tzu said: If you know the ___(i)___ and know ___(ii)___, your victory will not stand in doubt; if you know ___(iii)___ and know ___(iv)___, you may make your victory complete.

	(i)	(ii)	(iii)	(iv)
(a)	Heaven	yourself	enemy	Earth
(b)	enemy	Earth	yourself	Heaven
(c)	Earth	enemy	Heaven	yourself
(d)	enemy	yourself	Heaven	Earth

10

TERRAIN

Answers

Q1. (b)
Q2. (a), (b), (c)
Q3. (a), (b), (c)
Q4. (c)
Q5. (d)
Q6. (c)
Q7. (c)
Q8. (a), (c)
Q9. (d)
Q10. True
Q11. (d)
Q12. True
Q13. (b)
Q14. (c)
Q15. (a), (b), (c), (d)
Q16. (d)
Q17. (b)
Q18. (c)
Q19. (d)
Q20. True
Q21. (d)
Q22. (d)
Q23. (b)
Q24. (c)
Q25. (d)

11
THE NINE SITUATIONS

Instructions:
1. Select the appropriate answer(s) from the questions with four options (a),
(b), (c) and (d). In some questions there can be more than one correct
answer.
2. In the questions with True or False options, choose the appropriate
option.

Q1. Sun Tzu said: The art of war recognizes nine varieties of ground: (1)
Dispersive ground; (2) facile ground; (3) contentious ground; (4) ___(i)___;
(5) ground of intersecting highways; (6) ___(ii)___; (7) difficult ground; (8)
hemmed-in ground; (9) ___(iii)___.

(i)	(ii)	(iii)
(a) open ground	entangling ground	serious ground
(b) serious ground	accessible ground	open ground
(c) open ground	serious ground	desperate ground
(d) desperate ground	open ground	entangling ground

Q2. Which of the following is (are) true of dispersive ground?

(a) When a chieftain is fighting in his own territory, it is dispersive ground.
(b) On dispersive ground, fight not.
(c) On dispersive ground, chieftain should inspire his men with unity of
purpose.

(d) When chieftain has penetrated into hostile territory, but to no great distance, it is dispersive ground.

Q3. Which of the following is (are) true of facile ground?

(a) On facile ground, halt not.
(b) When chieftain has penetrated into hostile territory, but to no great distance, it is facile ground.
(c) Ground on which each side has liberty of movement is facile ground.
(d) On facile ground, chieftain would see that there is close connection between all parts of my army.

Q4. Which of the following is (are) true of contentious ground?

(a) On contentious ground, attack not.
(b) On contentious ground join hands with your allies.
(c) Ground the possession of which imports great advantage to either side, is contentious ground.
(d) On contentious ground, keep steadily on the march.

Q5. Which of the following is (are) NOT true of open ground?

(a) Ground on which each side has liberty of movement is open ground.
(b) On open ground, attack the enemy.
(c) On open ground, do not try to block the enemy's way.
(d) On open ground, keep a vigilant eye on your defenses.

Q6. Which of the following is (are) NOT true of ground of intersecting highways?

(a) Ground which forms the key to three contiguous states, so that he who occupies it first has most of the Empire at his command, is a ground of intersecting highways.
(b) On ground of intersecting highways, fight without delay.
(c) On ground of intersecting highways, join hands with allies.

(d) When there are means of communication on all four sides, the ground is one of intersecting highways.

Q7. Which of the following is (are) true of serious ground?
(a) On serious ground, try to ensure a continuous stream of supplies.
(b) On serious ground, gather in plunder.
(c) On serious ground don't try to block the enemy's way.
(d) When an army has penetrated into the heart of a hostile country, leaving a number of fortified cities in its rear, it is serious ground.

Q8. Which of the following is (are) NOT true of difficult ground?

(a) In difficult ground, keep steadily on the march.
(b) In difficult ground join hands with your allies.
(c) Mountain forests, rugged steeps, marshes and fens - all country that is hard to traverse: this is difficult ground.
(d) On difficult ground, don't encamp and keep pushing on along the road.

Q9. Which of the following is (are) true of hemmed-in ground?

(a) On hemmed-in ground, resort to stratagem.
(b) When you have the enemy's strongholds on your rear, and narrow passes in front, it is hemmed-in ground.
(c) When there is no place of refuge at all, it is hemmed-in ground.
(d) Ground which is reached through narrow gorges, and from which we can only retire by tortuous paths, so that a small number of the enemy would suffice to crush a large body of our men: this is hemmed in ground.

Q10. Which of the following is (are) NOT true of desperate ground?

(a) Ground on which we can only be saved from destruction by fighting without delay, is desperate ground.
(b) On desperate ground, fight not.
(c) On desperate ground proclaim to your soldiers the hopelessness of saving their lives.
(d) When there is no place of refuge at all, it is desperate ground.

Q11. Sun Tzu said: Those who were called skillful leaders of old knew how to -

(a) drive a wedge between the enemy's front and rear
(b) prevent co-operation between his large and small divisions
(c) hinder the good troops from rescuing the bad
(d) hinder the officers from rallying their men

Q12. Which of the following is (are) true of skillful leaders of old ?

(a) When the enemy's men were united, they managed to keep them in disorder.
(b) When it was to their advantage, they made a forward move.
(c) When the enemy's men were united, they managed to keep them in order.
(d) When it was to their disadvantage, they stopped still.

Q13. Sun Tzu said: If asked how to cope with a great host of the ___(i)___ in orderly array and on the point of marching to the ___(ii)___, I should say: "Begin by seizing something which your opponent holds ___(iii)___; then he will be ___(iv)___ to your will."

	(i)	(ii)	(iii)	(iv)
(a)	attack	enemy	amenable	dear
(b)	enemy	dear	attack	dear
(c)	attack	enemy	dear	amenable
(d)	enemy	attack	dear	amenable

Q14. Sun Tzu said: ___(i)___ is the essence of war: take advantage of the enemy's ___(ii)___, make your way by ___(iii)___ routes, and attack ___(iv)___ spots.

	(i)	(ii)	(iii)	(iv)
(a)	unexpected	rapidity	unguarded	unreadiness
(b)	unguarded	unreadiness	rapidity	unexpected
(c)	rapidity	unreadiness	unexpected	unguarded
(d)	unexpected	unreadiness	rapidity	unguarded

Q15. Which of the following is (are) the principles to be observed by an invading force?

(a) Make forays in fertile country in order to supply your army with food.
(b) The further you penetrate into a country, the greater will be the solidarity of your troops, and thus the defenders will not prevail against you.
(c) Carefully study the well-being of your men, and do not overtax them. Keep your army continually on the move, and devise unfathomable plans.
(d) Throw your soldiers into positions whence there is no escape, and they will prefer death to flight. If they will face death, there is nothing they may not achieve.

Q16. Sun Tzu said: Soldiers when in ___(i)___ straits lose the sense of ___(ii)___. If there is no place of refuge, they will stand firm. If they are in ___(iii)___ country, they will show a ___(iv)___ front. If there is no help for it, they will fight hard.

(i)	(ii)	(iii)	(iv)
(a) hostile	desperate	stubborn	fear
(b) desperate	fear	hostile	stubborn
(c) hostile	fear	desperate	stubborn
(d) desperate	stubborn	hostile	fear

Q17. Sun Tzu said: Prohibit the taking of ___(i)___, and do away with ___(ii)___ doubts. Then, until death itself comes, no ___(iii)___ need be ___(iv)___.

(i)	(ii)	(iii)	(iv)
(a) superstitious	feared	omens	calamity
(b) feared	superstitious	calamity	omens
(c) omens	superstitious	calamity	feared
(d) calamity	feared	omens	superstitious

Q18. Sun Tzu said: The principle on which to manage an army is to set up one standard of courage which all must reach. (True or False)

Q19. Which of the following is (are) true of the skillful general?

(a) The skillful general conducts his army just as though he were leading a single man, willy-nilly, by the hand.
(b) It is the business of a general to be quiet and thus ensure secrecy; upright and just, and thus maintain order.
(c) He must be able to mystify his officers and men by false reports and appearances, and thus keep them in total ignorance.
(d) By altering his arrangements and changing his plans, he keeps the enemy without definite knowledge. By shifting his camp and taking circuitous routes, he prevents the enemy from anticipating his purpose.

Q20. At the critical moment, how does the leader of an army acts?

(a) He acts like one who has climbed up a height and then kicks away the ladder behind him.
(b) He carries his men deep into hostile territory before he shows his hand.
(c) He burns his boats and breaks his cooking-pots.
(d) Like a shepherd driving a flock of sheep, he drives his men this way and that, and nothing knows whither he is going.

Q21. Which of the following things that must most certainly be studied?

(a) The different measures suited to the nine varieties of ground.
(b) The expediency of aggressive or defensive tactics.
(c) The fundamental laws of human nature.
(d) None of these.

Q22. Sun Tzu said: When invading hostile territory, the general principle is, that penetrating ___(i)___ brings ___(ii)___; penetrating but a ___(iii)___ way means ___(iv)___.

(i)	(ii)	(iii)	(iv)
(a) short	cohesion	dispersion	deeply
(b) cohesion	dispersion	short	deeply
(c) dispersion	deeply	short	cohesion

(d) deeply   cohesion   short   dispersion

Q23. Sun Tzu said: It is the soldier's disposition to offer an obstinate ___(i)___ when ___(ii)___, to fight hard when he cannot help himself, and to ___(iii)___ promptly when he has fallen into ___(iv)___.

(i)	(ii)	(iii)	(iv)
(a) danger	resistance	surrounded	obey
(b) obey	danger	surrounded	resistance
(c) resistance	surrounded	obey	danger
(d) danger	surrounded	obey	resistance

Q24. Sun Tzu said: We cannot enter into alliance with neighboring princes until we are acquainted with their designs. (True or False)

Q25. Sun Tzu said: We are fit to lead an army on the march unless we are unfamiliar with the face of the country - its mountains and forests, its pitfalls and precipices, its marshes and swamps. (True or False)

Q26. Sun Tzu said: We shall be unable to turn natural advantages to account unless we make use of local guides. (True or False)

Q27. Which of the following is (are) true of warlike prince?

(a) When a warlike prince attacks a powerful state, his generalship shows itself in preventing the concentration of the enemy's forces.
(b) He overawes his opponents, and their allies are prevented from joining against him.
(c) He does not strive to ally himself with all and sundry, nor does he foster the power of other states.
(d) He carries out his own secret designs, keeping his antagonists in awe. Thus he is able to capture their cities and overthrow their kingdoms.

Q28. Sun Tzu said: Confront your soldiers with the ___(i)___ itself; never let them know your ___(ii)___. When the outlook is ___(ii)___, bring it before their eyes; but tell them nothing when the situation is ___(iv)___.

	(i)	(ii)	(iii)	(iv)
(a)	gloomy	bright	deed	design
(b)	design	deed	gloomy	bright
(c)	bright	design	gloomy	deed
(d)	deed	design	bright	gloomy

Q29. Sun Tzu said: Place your army in deadly ___(i)___, and it will ___(ii)___; plunge it into desperate ___(iii)___, and it will come off in ___(iv)___. For it is precisely when a force has fallen into harm's way that is capable of striking a blow for victory.

	(i)	(ii)	(iii)	(iv)
(a)	straits	survive	safety	peril
(b)	safety	peril	straits	survive
(c)	straits	survive	peril	straits
(d)	peril	survive	straits	safety

Q30. Sun Tzu said: Success in warfare is gained by carefully accommodating ourselves to the enemy's purpose. (True or False)

Q31. Sun Tzu said: Forestall your opponent by ___(i)___ what he holds ___(ii)___, and subtly contrive to ___(iii)___ his ___(iv)___ on the ground.

	(i)	(ii)	(iii)	(iv)
(a)	time	arrival	seizing	dear
(b)	arrival	dear	time	seizing
(c)	seizing	dear	time	arrival
(d)	dear	seizing	time	arrival

Q32. Sun Tzu said: Walk in the path defined by rule, and accommodate yourself to the enemy until you can fight a decisive battle. (True or False)

Q33. Sun Tzu said: At first, then, exhibit the coyness of a maiden, until the ___(i)___ gives you an ___(ii)___; afterwards emulate the rapidity of a running hare, and it will be too ___(iii)___ for the enemy to ___(iv)___ you.

(i)	(ii)	(iii)	(iv)
(a) late	enemy	oppose	opening
(b) enemy	late	opening	oppose
(c) opening	enemy	oppose	late
(d) enemy	opening	late	oppose

11

THE NINE SITUATIONS

Answers

Q1. (c)

Q2. (a), (b), (c)

Q3. (a), (b), (d)

Q4. (a), (c)

Q5. (b)

Q6. (b)

Q7. (a), (b), (d)

Q8. (b)

Q9. (a), (b), (d)

Q10. (b)

Q11. (a), (b), (c), (d)

Q12. (a), (b), (d)

Q13. (d)

Q14. (c)

Q15. (a), (b), (c), (d)

Q16. (b)

Q17. (c)

Q18. True

Q19. (a), (b), (c)

Q20. (a), (b), (c), (d)

Q21. (a), (b), (c)

Q22. (d)

Q23. (c)

Q24. True

Q25. False

Q26. True

Q27. (a), (b), (c), (d)

Q28. (d)

Q29. (d)

Q30. True

Q31. (c)

Q32. True

Q33. (d)

12
THE ATTACK BY FIRE

Instructions:
1. Select the appropriate answer(s) from the questions with four options (a), (b), (c) and (d). In some questions there can be more than one correct answer.
2. In the questions with True or False options, choose the appropriate option.

Q1. Sun Tzu said: There are five ways of attacking with fire.
Which of the following is (are) part of those five ways?

(a) Burn soldiers in their camp
(b) Burn stores and burn baggage trains
(c) Burn arsenals and magazines
(d) Hurl dropping fire amongst the enemy

Q2. Sun Tzu said: In order to carry out an ___(i)___, we must have ___(ii)___ available. The material for raising ___(iii)___ should always be kept in ___(iv)___.

(i)	(ii)	(iii)	(iv)
(a) fire	readiness	attack	means
(b) fire	means	attack	readiness
(c) attack	means	fire	readiness
(d) attack	readiness	fire	means

Q3. Sun Tzu said: There is a proper season for making attacks with ___(i)___, and special days for starting a ___(ii)___. The proper season is when the weather is ___(iii)___; the special days are those when the moon is in the constellations of the Sieve, the Wall, the Wing or the Cross-bar; for these four are all days of rising ___(iv)___.

(i)	(ii)	(iii)	(iv)
(a) wind	fire	conflagration	very dry
(b) conflagration	wind	very dry	fire
(c) fire	conflagration	very dry	wind
(d) wind	fire	very dry	conflagration

Q4. Sun Tzu said: In attacking with fire, when fire breaks out ___(i)___ to enemy's ___(ii)___, respond at once with an ___(iii)___ from ___(iv)___.

(i)	(ii)	(iii)	(iv)
(a) camp	attack	inside	without
(b) without	camp	attack	inside
(c) attack	inside	without	camp
(d) inside	camp	attack	without

Q5. Sun Tzu said: In attacking with fire, if there is an outbreak of fire, but the enemy's ___(i)___ remain ___(ii)___, bide your ___(iii)___ and do not ___(iv)___.

(i)	(ii)	(iii)	(iv)
(a) attack	quiet	soldiers	time
(b) time	attack	soldiers	quiet
(c) quiet	soldiers	attack	time
(d) soldiers	quiet	time	attack

Q6. Sun Tzu said: In attacking with fire, when the force of the ___(i)___ has reached its ___(ii)___, follow it up with an ___(iii)___, if that is practicable; if not, ___(iv)___ where you are.

(i)	(ii)	(iii)	(iv)
(a) attack	height	flames	stay
(b) height	stay	flames	attack
(c) flames	height	attack	stay
(d) stay	flames	height	attack

Q7. Sun Tzu said: In attacking with fire, if it is possible to make an ___(i)___ with fire from without, do not wait for it to break out ___(ii)___, but deliver your ___(iii)___ at a ___(iv)___ moment.

(i)	(ii)	(iii)	(iv)
(a) favorable	attack	assault	within
(b) attack	assault	within	favorable
(c) assault	within	attack	favorable
(d) attack	favorable	within	assault

Q8. Sun Tzu said: In attacking with fire, when you ___(i)___ a fire, be to ___(ii)___ of it. Do not ___(iii)___ from the ___(iv)___.

(i)	(ii)	(iii)	(iv)
(a) attack	leeward	start	windward
(b) start	windward	attack	leeward
(c) windward	leeward	attack	start
(d) leeward	windward	start	attack

Q9. Sun Tzu said: A wind that rises in the ___(i)___ lasts ___(ii)___, but a ___(iii)___ breeze soon ___(iv)___.

(i)	(ii)	(iii)	(iv)
(a) night	daytime	long	falls
(b) falls	long	night	daytime
(c) long	night	daytime	falls
(d) daytime	long	night	falls

Q10. Sun Tzu said: In every army, the five developments connected with ___(i)___ must be known, the movements of the ___(ii)___ calculated, and a ___(iii)___ kept for the proper ___(iv)___.

	(i)	(ii)	(iii)	(iv)
(a)	stars	days	fire	watch
(b)	days	fire	watch	stars
(c)	fire	stars	watch	days
(d)	watch	days	fire	stars

Q11. Sun Tzu said: Those who use ___(i)___ as an aid to the attack show ___(ii)___; those who use ___(iii)___ as an aid to the attack gain an accession of ___(iv)___.

	(i)	(ii)	(iii)	(iv)
(a)	intelligence	strength	water	fire
(b)	water	intelligence	strength	fire
(c)	fire	intelligence	water	strength
(d)	strength	fire	intelligence	water

Q12. Sun Tzu said: By means of water, an enemy may be intercepted, but not robbed of all his belongings. (True or False)

Q13. Sun Tzu said: Unhappy is the fate of one who tries to win his battles and succeed in his ___(i)___ without cultivating the spirit of ___(ii)___; for the result is waste of ___(iii)___ and general ___(iv)___.

	(i)	(ii)	(iii)	(iv)
(a)	enterprise	time	attacks	stagnation
(b)	time	attacks	stagnation	enterprise
(c)	stagnation	time	enterprise	attacks
(d)	attacks	enterprise	time	stagnation

Q14. Sun Tzu said: The enlightened ___(i)___ lays his ___(ii)___ well ahead; the good general ___(iii)___ his ___(iv)___.

	(i)	(ii)	(iii)	(iv)
(a)	resources	cultivates	plans	ruler
(b)	plans	ruler	resources	cultivates
(c)	resources	plans	cultivates	ruler

(d) ruler   plans   cultivates   resources

Q15. Sun Tzu said: ___(i)___ not unless you see an ___(ii)___; use not your troops unless there is something to be gained; ___(iii)___ not unless the position is ___(iv)___.

(i)	(ii)	(iii)	(iv)
(a) fight	critical	move	advantage
(b) advantage	critical	move	fight
(c) move	advantage	fight	critical
(d) critical	fight	move	advantage

Q16. Sun Tzu said: No ___(i)___ should put troops into the field merely to gratify his own ___(ii)___; no ___(iii)___ should fight a battle simply out of ___(iv)___.

(i)	(ii)	(iii)	(iv)
(a) general	ruler	spleen	pique
(b) ruler	general	pique	spleen
(c) ruler	spleen	general	pique
(d) spleen	pique	ruler	general

Q17. Sun Tzu said: Anger may in time change to gladness; vexation may be succeeded by content. But a kingdom that has once been destroyed can never come again into being; nor can the dead ever be brought back to life. Hence the enlightened ruler is ___(i)___, and the good general full of ___(ii)___. This is the way to keep a country at ___(iii)___ and an army ___(iv)___.

(i)	(ii)	(iii)	(iv)
(a) intact	peace	caution	heedful
(b) peace	caution	intact	heedful
(c) caution	heedful	intact	peace
(d) heedful	caution	peace	intact

12

THE ATTACK BY FIRE

Answers

Q1. (a), (b), (c), (d)

Q2. (c)

Q3. (c)

Q4. (d)

Q5. (d)

Q6. (c)

Q7. (c)

Q8. (b)

Q9. (d)

Q10. (c)

Q11. (c)

Q12. True

Q13. (d)

Q14. (d)

Q15. (c)

Q16. (c)

Q17. (d)

13
THE USE OF SPIES

Instructions:
1. Select the appropriate answer(s) from the questions with four options (a), (b), (c) and (d). In some questions there can be more than one correct answer.
2. In the questions with True or False options, choose the appropriate option.

Q1. Sun Tzu said: ___(i)___ a host of a hundred thousand men and ___(ii)___ them great distances entails heavy loss on the ___(iii)___ and a drain on the ___(iv)___ of the State.

(i)	(ii)	(iii)	(iv)
(a) marching	raising	resources	people
(b) raising	people	marching	resources
(c) marching	raising	resources	people
(d) raising	marching	people	resources

Q2. Sun Tzu said: ___(i)___ armies may face each other for ___(ii)___, striving for the ___(iii)___ which is decided in a ___(iv)___ day.

(i)	(ii)	(iii)	(iv)
(a) single	victory	hostile	years
(b) victory	hostile	years	single
(c) hostile	years	victory	single
(d) years	hostile	single	victory

95

Q3. Sun Tzu said: To remain in ___(i)___ of the enemy's ___(ii)___ simply because one grudges the outlay of a hundred ounces of silver in honors and emoluments, is the height of inhumanity. One who acts thus is no ___(iii)___ of men, no present help to his sovereign, no master of ___(iv)___.

(i)	(ii)	(iii)	(iv)
(a) condition	victory	ignorance	leader
(b) victory	ignorance	leader	condition
(c) ignorance	condition	leader	victory
(d) leader	condition	victory	ignorance

Q4. Which of the following is (are) true of foreknowledge?

(a) Foreknowledge cannot be elicited from spirits.
(b) Foreknowledge cannot be obtained inductively from experience, nor by any deductive calculation.
(c) Foreknowledge enables the wise sovereign and the good general to strike and conquer, and achieve things beyond the reach of ordinary men.
(d) None of these.

Q5. Sun Tzu said: Knowledge of the enemy's dispositions can only be obtained from other men. Hence the use of spies. (True or False)

Q6. Sun Tzu said: The five classes of spies are : (1) Local spies; (2) ___(i)___; (3) converted spies; (4) ___(ii)___; (5) surviving spies.

(i)	(ii)
(a) outward spies	inverted spies
(b) inward spies	outward spies
(c) doomed spies	inverted spies
(d) inward spies	doomed spies

Q7. Sun Tzu said: When the five kinds of ___(i)___ are all at ___(ii)___, none can discover the ___(iii)___ system. This is called "divine manipulation of the threads." It is the sovereign's most ___(iv)___ faculty.

	(i)	(ii)	(iii)	(iv)
(a)	secret	spy	precious	work
(b)	precious	secret	spy	work
(c)	spy	work	secret	precious
(d)	work	spy	precious	secret

Q8. Which if the following is (are) true of spies ?
(a) Having local spies means making use of officials of the enemy.
(b) Having inward spies means employing the services of the inhabitants of a district.
(c) Having converted spies, getting hold of the enemy's spies and using them for our own purposes.
(d) Surviving spies those who doing certain things openly for purposes of deception

Q9. Which if the following is (are) NOT true of spies ?

(a) Having local spies means making use of officials of the enemy.
(b) Surviving spies are those who bring back news from the enemy's camp.
(c) Having inward spies means employing the services of the inhabitants of a district.
(d) Doomed spies knowingly do certain things openly for purposes of deception, and report them to the enemy.

Q10. Sun Tzu said: Having ___(i)___ spies means employing the services of the inhabitants of a district. Having ___(ii)___ spies, making use of officials of the enemy. Having ___(iii)___ spies, getting hold of the enemy's spies and using them for our own purposes. ___(iv)___ spies, finally, are those who bring back news from the enemy's camp.

	(i)	(ii)	(iii)	(iv)
(a)	local	converted	inward	surviving
(b)	inward	local	surviving	converted
(c)	surviving	converted	local	inward
(d)	local	inward	converted	surviving

Q11. Which of the following is (are) true of spies ?

(a) Spies cannot be usefully employed without a certain intuitive sagacity.
(b) None should be more liberally rewarded than the spies. None in the whole army are more intimate relations to be maintained than with spies.
(c) Spies can be properly managed without benevolence and straightforwardness.
(d) Without subtle ingenuity of mind, one cannot make certain of the truth of their reports.

Q12. Sun Tzu said: If a secret piece of news is ___(i)___ by a spy before the time is ___(ii)___, he must be put to ___(iii)___ together with the man to whom the secret was ___(iv)___.

(i)	(ii)	(iii)	(iv)
(a) told	divulged	ripe	death
(b) ripe	told	death	divulged
(c) divulged	ripe	death	told
(d) told	death	ripe	divulged

Q13. Sun Tzu said: Whether the object be to crush an army, to storm a city, or to assassinate an individual, it is always necessary to begin by ___(i)___ out the ___(ii)___ of the attendants, the aides-de-camp, and door-keepers and sentries of the general in command. Our ___(iii)___ must be commissioned to ___(iv)___ these.

(i)	(ii)	(iii)	(iv)
(a) ascertain	spies	names	finding
(b) finding	spies	names	ascertain
(c) ascertain	finding	spies	names
(d) finding	names	spies	ascertain

Q14. Sun Tzu said: The enemy's spies who have come to spy on us must be sought out, tempted with bribes, led away and comfortably housed. Thus they will become _____ spies and available for our service.

(a) inward
(b) converted
(c) doomed
(d) surviving

Q15. Sun Tzu said: It is essential that the converted spy be treated with the utmost liberality. Why ?

(a) It is through the information brought by the converted spy that we are able to acquire and employ local and inward spies.
(b) It is owing to the information brought by the converted spy, that we can cause the doomed spy to carry false tidings to the enemy.
(c) It is by information brought by the converted spy that the surviving spy can be used on appointed occasions.
(d) The end and aim of spying in all its five varieties is knowledge of the enemy; and this knowledge can only be derived, in the first instance, from the converted spy.

Q16. Sun Tzu said: It is only the enlightened ruler and the wise ___(i)___ who will use the highest ___(ii)___ of the army for purposes of ___(iii)___ and thereby they achieve great ___(iv)___ .

(i)	(ii)	(iii)	(iv)
(a) spying	results	intelligence	general
(b) intelligence	spying	general	results
(c) general	intelligence	spying	results
(d) results	general	intelligence	spying

Q17. Sun Tzu said: ___(i)___ are a most important element in ___(ii)___, because on them depends an ___(iii)___ ability to ___(iv)___ .

(i)	(ii)	(iii)	(iv)
(a) army's	war	spies	move
(b) war	spies	army's	move
(c) army's	move	spies	war
(d) spies	war	army's	move

13

THE USE OF SPIES

Answers

Q1. (d)
Q2. (c)
Q3. (c)
Q4. (a), (b), (c)
Q5. True
Q6. (d)
Q7. (c)
Q8. (c)
Q9. (a), (c)
Q10. (d)
Q11. (a), (b), (d)
Q12. (c)
Q13. (d)
Q14. (b)
Q15. (a), (b), (c), (d)
Q16. (c)
Q17. (d)

PART II

In this part the questions and answers on each chapter of *The Art of War* are together. The correct answers are in **bold** typeface and the wrong ones are in normal typeface.

1
LAYING PLANS

Q1. Sun Tzu said: The art of war is of vital importance to the State. It is a matter of ___(i)___ and ___(ii)___, a road either to ___(iii)___ or to ___(iv)___. Hence it is a subject of inquiry which can on no account be neglected.

(i)	(ii)	(iii)	(iv)
(a) life	safety	death	ruin
(b) safety	ruin	life	death
(c) life	**death**	**safety**	**ruin**
(c) life	ruin	safety	death

Q2. What are the five constant factors that govern the art of war ?

(a) The Moral Law; Heaven; Sea; The Commander; Method and Discipline
(b) The Moral Law; Heaven; Earth; The Commander; The State
(c) The Moral Law; Heaven; Earth; The Commander; Method and Discipline
(d) The Moral Law; Heaven; Earth; Sea; The Commander

Q3. What does the Moral Law, a constant factor that governs the art of war, causes the people to be?

(a) It causes the people to be in complete discord with their enemy.
(b) It causes the people to be in complete accord with their ruler.
(c) It causes the ruler to be in complete accord with the people.

(d) It causes the people to be in complete accord with themselves.

Q4. What does Heaven, one of the five constant factors that govern the art of war, do not signify?

(a) day and night
(b) cold and heat
(c) danger and security
(d) times and seasons

Q5. What does the Earth, one of the five constant factors that govern the art of war, comprises of?

(a) danger and security
(b) distances great and small
(c) open ground and narrow passes
(d) the chances of life and death

Q6. The Commander, a constant factor that governs the art of war, stands for virtues of -

(a) wisdom and sincerity only
(b) wisdom, sincerity and benevolence only
(c) courage and strictness only
(d) wisdom, sincerity, benevolence, courage and strictness

Q7. Sun Tzu said: By Method and Discipline are to be understood the marshaling of the army in its proper ___(i)___, the graduations of rank among the ___(ii)___, the ___(iii)___ of roads by which supplies may reach the army, and the control of military ___(iv)___.

(i)	(ii)	(iii)	(iv)
(a) officers	subdivisions	expenditure	maintenance
(b) expenditure	subdivisions	maintenance	officers
(c) maintenance	expenditure	subdivisions	officers
(d) subdivisions	**officers**	**maintenance**	**expenditure**

Q8. Sun Tzu said: The general who knows the ___(i)___ constant factors which govern the art of war, will be ___(ii)___; he who knows them not will ___(iii)___.

(i)	(ii)	(iii)
(a) three	victorious	fail
(b) four	fail	victorious
(c) five	**victorious**	**fail**
(d) six	fail	victorious

Q9. In your deliberations, when seeking to determine the military conditions, which of the following can be the basis of a comparison?

(a) Which of the two sovereigns is imbued with the Moral law?
(b) Which of the two generals has most ability?
(c) With whom lie the advantages derived from Heaven and Earth?
(d) On which side is discipline most rigorously enforced?

Q10. By means of which considerations, Sun Tzu can forecast victory or defeat?

(a) (1) Which of the two sovereigns is imbued with the Moral law?
    (2) Which of the two generals has most ability?
(b) (1) With whom lie the advantages derived from Heaven and Earth?
    (2) On which side is discipline most rigorously enforced?
(c) (1) Which army is stronger?
    (2) On which side are officers and men more highly trained?
(d) In which army is there the greater constancy both in reward and punishment?

Q11. Sun Tzu said: The general that hearkens to my counsel and acts upon it, will ___(i)___ :- let such a one be ___(ii)___ in command! The general that hearkens not to my counsel nor acts upon it, will suffer ___(iii)___ :- let such a one be ___(iv)___!

(i)	(ii)	(iii)	(iv)
(a) conquer	dismissed	defeat	retained
(b) defeat	retained	conquer	dismissed
(c) conquer	**retained**	**defeat**	**dismissed**
(d) defeat	conquer	dismissed	retained

Q12. Sun Tzu said: While heading the profit of my counsel, avail yourself also of any helpful ___(i)___ over and beyond the ___(ii)___ rules. According as circumstances are ___(iii)___, one should modify one's ___(iv)___.

(i)	(ii)	(iii)	(iv)
(a) plans	favorable	ordinary	circumstances
(b) ordinary	favorable	plans	circumstances
(c) circumstances	**ordinary**	**favorable**	**plans**
(d) plans	circumstances	ordinary	favorable

13. Sun Tzu said: All warfare is based on deception. Hence -

(a) when able to attack, we must seem unable.
(b) when using our forces, we must seem inactive.
(c) when we are near, we must make the enemy believe we are far away.
(d) when far away, we must make enemy believe we are near.

Q14. Sun Tzu said: These military devices, leading to victory, must not be divulged beforehand –

(a) Hold out baits to entice the enemy. Feign disorder, and crush him.
(b) If he is not secure at all points, be prepared for him. If he is in superior strength, evade him.
(c) If he is taking his ease, give him no rest. If his forces are united, separate them.
(d) Attack enemy where he is unprepared, appear where you are expected.

Q15. Sun Tzu said: If your opponent is of choleric temper, seek to ___(i)___ him. Pretend to be ___(ii)___, that he may grow ___(iii)___.

	(i)	(ii)	(iii)
(a)	weak	arrogant	irritate
(b)	irritate	arrogant	weak
(c)	weak	irritate	weak
(d)	**irritate**	**weak**	**arrogant**

Q16. Sun Tzu said: The general who ___(i)___ a battle makes ___(ii)___ calculations in his temple before the battle is fought. The general who ___(iii)___ a battle makes but ___(iv)___ calculations beforehand.

	(i)	(ii)	(iii)	(iv)
(a)	wins	few	loses	many
(b)	loses	many	wins	few
(c)	**wins**	**many**	**loses**	**few**
(d)	loses	few	wins	many

107

2
WAGING WAR

Q1. Sun Tzu said: When you engage in actual fighting, if ___(i)___ is long in coming, then men's weapons will grow ___(ii)___ and their ardor will be ___(iii)___. If you lay siege to a town, you will ___(iv)___ your strength.

(i)	(ii)	(iii)	(iv)
(a) victory	exhaust	damped	dull
(b) exhaust	dull	victory	damped
(c) damped	exhaust	dull	victory
(d) victory	**dull**	**damped**	**exhaust**

Q2. Sun Tzu said: When you engage in actual fighting, if the campaign is ___(i)___, the resources of the ___(ii)___ will not be equal to the ___(iii)___.

(i)	(ii)	(iii)
(a) strain	State	protracted
(b) State	protracted	strain
(c) protracted	**State**	**strain**
(d) strain	protracted	State

Q3. Sun Tzu said: In a prolonged warfare, when your weapons are dulled, your ardor damped, your strength exhausted and your treasure spent, other chieftains will spring up to take advantage of your extremity. Then no man, however wise, will be able to avert the consequences that must ensue. (**True** or False)

Q4. Sun Tzu said: There is no instance of a country having benefited from prolonged warfare. (**True** or False)

Q5. Sun Tzu said: It is only one ___(i)___ is thoroughly ___(ii)___ with the ___(iii)___ of war that can thoroughly understand the ___(iv)___ way of carrying it on.

(i)	(ii)	(iii)	(iv)
(a) acquainted	profitable	who	evils
(b) evils	profitable	acquainted	who
(c) who	**acquainted**	**evils**	**profitable**
(d) profitable	acquainted	who	evils

Q6. Sun Tzu said: The skillful soldier does not raise a second levy, neither are his supply-wagons loaded more than twice. (**True** or False)

Q7. Sun Tzu said: Bring war material with you from home, but forage on the enemy. Thus the army will have food enough for its needs. (**True** or False)

Q8. A wise general makes a point of foraging on the enemy, why?

(a) Contributing to maintain an army at a distance causes the people to be impoverished.
(b) The proximity of an army causes prices to go up; and high prices cause the people's substance to be drained away.
(c) When people's substance is drained away, the peasantry will be afflicted by heavy exactions.
(d) One cartload of the enemy's provisions is equivalent to twenty of one's
own.

Q9. Sun Tzu said: In order to kill the ___(i)___, our men must be roused to ___(ii)___; that there may be ___(iii)___ from defeating the enemy, they must have their ___(iv)___.

(i)	(ii)	(iii)	(iv)
(a) anger	enemy	rewards	advantage
(b) advantage	rewards	anger	enemy
(c) enemy	**anger**	**advantage**	**rewards**
(d) rewards	enemy	anger	advantage

Q10. Which of the following is (are) using the conquered foe to augment one's own strength ?

(a) The captured soldiers kindly treated and kept.
(b) Our own flags substituted for those of the enemy on chariots taken in chariot fighting, and the chariots mingled and used in conjunction with ours.
(c) Bringing war material with you from home, but foraging on the enemy.
(d) None of these.

Q11. Sun Tzu said: In war let your great object be victory, not lengthy campaigns. (**True** or False)

Q12. Sun Tzu said: The ___(i)___ of armies is the arbiter of the people's ___(ii)___, the man on whom it depends whether the ___(iii)___ shall be in ___(iv)___ or in peril.

(i)	(ii)	(iii)	(iv)
(a) fate	nation	peace	leader
(b) peace	rewards	nation	fate
(c) nation	fate	leader	peace
(d) leader	**fate**	**nation**	**peace**

3
ATTACK BY STRATAGEM

Q1. Sun Tzu said: In the practical art of war, the best thing of all is -

(a) to shatter and destroy the enemy's country.
(b) to recapture an army entire than to destroy it.
(c) to take the enemy's country whole and intact.
(d) to capture a regiment, a detachment or a company entire than to destroy them.

Q2. Sun Tzu said: In the art of war, the supreme excellence is -

(a) to fight and conquer in all your battles.
(b) breaking the enemy's resistance without fighting.
(c) to fight and win no battle.
(d) to shatter and destroy the enemy's country.

Q3. Arrange the following four forms of generalship from the highest form to the lowest form.

(I) to attack the enemy's army in the field
(II) to besiege walled cities
(III) to balk the enemy's plans
(IV) to prevent the junction of the enemy's forces

(a) (I), (II), (III), (IV)
(b) (III), (I), (II), (IV)
(c) (II), (IV), (I), (III)
(d) (III), (IV), (I), (II)

Q4. Sun Tzu said: The rule is, not to besiege walled cities if it can possibly be avoided. Why?

(a) The preparation of mantlets, movable shelters, and various implements of war, will take up three whole months.
(b) The piling up of mounds over against the walls will take three months.
(c) The general, unable to control his irritation, will launch his men to the assault like swarming ants, with the result that one-third of his men are slain, while the town still remains untaken. Such are the disastrous effects of a siege.
(d) None of these.

Q5. In war, what does skillful leader do ?

(a) The skillful leader subdues the enemy's troops without any fighting.
(b) He captures enemy cities without laying siege to them.
(c) He overthrows enemy kingdom without lengthy operations in the field.
(d) With his forces intact he disputes the mastery of the Empire, and thus, without losing a man, he completes triumph.

Q6. Sun Tzu said: It is the rule in war, if our forces are ten to the enemy's one, then

(a) attack the enemy
(b) offer battle to the enemy
(c) surround the enemy
(d) avoid the enemy

Q7. Sun Tzu said: It is the rule in war, if our forces are five to the enemy's one then attack the enemy. (**True** or False)

Q8. Sun Tzu said: It is the rule in war, if our forces are twice as numerous as the enemy, then

(a) attack the enemy
(b) offer battle to the enemy
(c) avoid the enemy
(d) divide our army into two

Q9. Sun Tzu said: It is the rule in war, if our forces are equally matched in numbers to the enemy's, we can ___(i)___; if slightly inferior in numbers, we can ___(ii)___ the enemy; if quite unequal in every way, we can ___(iii)___ from the enemy.

(i)	(ii)	(iii)
(a) avoid	offer battle	flee
(b) avoid	flee	offer battle
(c) offer battle	**avoid**	**flee**
(d) flee	offer battle	avoid

10. Sun Tzu said: The ___(i)___ is the bulwark of the ___(ii)___; if the bulwark is complete at all points; the State will be ___(iii)___; if the bulwark is defective, the State will be ___(iv)___.

(i)	(ii)	(iii)	(iv)
(a) State	strong	general	weak
(b) strong	State	weak	general
(c) general	**State**	**strong**	**weak**
(d) State	general	weak	strong

Q11. The way(s) in which a ruler can bring misfortune upon his army is (are) -

(a) By commanding the army to advance or to retreat, being ignorant of the fact that it cannot obey.
(b) By attempting to govern an army in the same way as he administers a kingdom, being ignorant of the conditions which obtain in an army.
(c) By employing the officers of his army without discrimination, through ignorance of the military principle of adaptation to circumstances.
(d) None of these.

Q12. In art of war, which of the following is (are) part of the five essentials for victory?

(a) He will win who knows when to fight and when not to fight.
(b) He will win who knows how to handle both superior and inferior forces.
(c) He will win who, prepared himself, waits to take the enemy unprepared.

(d) He will win whose army is animated by the same spirit throughout all its ranks.

Q13. Sun Tzu said: If you know the enemy and know yourself then

(a) for every victory gained you will also suffer a defeat.
(b) you will succumb in every battle.
(c) you need not fear the result of a hundred battles.
(d) you can't break the enemy's resistance without fighting.

Q14. Sun Tzu said: If you know yourself but not the enemy then

(a) you will succumb in every battle.
(b) you need not fear the result of a hundred battles.
(c) for every victory gained you will also suffer a defeat.
(d) you will break the enemy's resistance without fighting.

Q15. Sun Tzu said: If you know neither the enemy nor yourself then

(a) you need not fear the result of a hundred battles.
(b) you will break the enemy's resistance without fighting.
(c) for every victory gained you will also suffer a defeat.
(d) you will succumb in every battle.

4

TACTICAL DISPOSITIONS

Q1. Sun Tzu said: The good ___(i)___ of old first put themselves beyond the possibility of ___(ii)___, and then ___(iii)___ for an opportunity of defeating the ___(iv)___.

(i)	(ii)	(iii)	(iv)
(a) enemy	fighters	defeat	waited
(b) fighters	enemy	waited	defeat
(c) waited	fighters	enemy	defeat
(d) fighters	**defeat**	**waited**	**enemy**

Q2. Sun Tzu said: To secure ourselves against defeat lies in our own hands, but -

(a) the opportunity of defeating the enemy is provided by the enemy ourselves.

(b) the opportunity of defeating the enemy is not provided by the enemy himself.

(c) the opportunity of defeating the enemy is provided by the enemy himself.

(d) the opportunity of defeating ourselves is provided by the enemy.

Q3. Sun Tzu said: The good fighter is able to secure himself against defeat, but cannot make certain of defeating the enemy. (**True** or False)

115

Q4. Sun Tzu said: Security against defeat implies ___(i)___ tactics; ability to defeat the enemy means taking the ___(ii)___. Standing on the defensive indicates insufficient ___(iii)___; ___(iv)___, a superabundance of strength.

	(i)	(ii)	(iii)	(iv)
(a)	offensive	defensive	attacking	strength
(b)	defensive	defensive	strength	attacking
(c)	offensive	offensive	attacking	strength
(d)	**defensive**	**offensive**	**strength**	**attacking**

Q5. Sun Tzu said: The general who is skilled in ___(i)___ hides in the most secret recesses of the earth; he who is skilled in ___(ii)___ flashes forth from the topmost heights of heaven. Thus on the one hand we have ability to ___(iii)___ ourselves; on the other, a ___(iv)___ that is complete.

	(i)	(ii)	(iii)	(iv)
(a)	attack	defense	protect	victory
(b)	victory	attack	protect	defense
(c)	protect	defense	victory	attack
(d)	**defense**	**attack**	**protect**	**victory**

Q6. Sun Tzu said: To see victory only when it is within the ken of the common herd is the acme of excellence. (True or **False**)

Q7. Sun Tzu said: It is not the acme of excellence if you fight and conquer and the whole Empire says, "Well done!" (**True** or False)

Q8. Sun Tzu said: What the ancients called a clever fighter is the one -

(a) who wins his battles by making no mistakes.
(b) who not only wins, but excels in winning with ease.
(c) whose victories bring him neither reputation for wisdom nor credit for courage.
(d) who puts himself into a position which makes defeat impossible, and does not miss the moment for defeating the enemy.

Q9. Sun Tzu said: In war the ___(i)___ strategist only seeks battle ___(ii)___ the victory has been won, whereas he who is destined to ___(iii)___ ___(iv)___ fights and afterwards looks for victory.

(i)	(ii)	(iii)	(iv)
(a) defeat	before	victorious	after
(b) victorious	before	defeat	first
(c) defeat	after	victorious	first
(d) victorious	**after**	**defeat**	**first**

Q10. Sun Tzu said: The consummate ___(i)___ cultivates the ___(ii)___, and strictly adheres to method and ___(iii)___; thus it is in his power to ___(iv)___ success.

(i)	(ii)	(iii)	(iv)
(a) discipline	control	moral law	leader
(b) moral law	discipline	leader	control
(c) leader	**moral law**	**discipline**	**control**
(d) control	leader	moral law	discipline

Q11. Sun Tzu said: In respect of military method, we have, firstly, Measurement; secondly, Estimation of quantity; thirdly, Calculation; fourthly, Balancing of chances; fifthly, Victory. Measurement owes its existence to ___(i)___; Estimation of quantity to ___(ii)___; Calculation to Estimation of quantity; Balancing of chances to ___(iii)___; and ___(iv)___ to Balancing of chances.

(i)	(ii)	(iii)	(iv)
(a) Victory	Calculation	Measurement	Earth
(b) Calculation	Victory	Earth	Measurement
(c) Earth	**Measurement**	**Calculation**	**Victory**
(d) Measurement	Calculation	Victory	Earth

Q12. Sun Tzu said: A ___(i)___ army opposed to a ___(ii)___ one, is as a pound's weight placed in the scale against a single grain. The ___(iii)___ of

117

a conquering ___(iv)___ is like the bursting of pent-up waters into a chasm a thousand fathoms deep.

(i)	(ii)	(iii)	(iv)
(a) routed	onrush	force	victorious
(b) onrush	victorious	routed	victorious
(c) victorious	**routed**	**onrush**	**force**
(d) routed	victorious	force	onrush

5
ENERGY

Q1. Sun Tzu said: The control of a large force is the same principle as the control of a few men: it is merely a question of -

(a) instituting signs and signals.
(b) direct and indirect maneuvers.
(c) science of weak points and strong.
(d) dividing up their numbers.

Q2. Sun Tzu said: Fighting with a large army under your command is nowise different from fighting with a small one: it is merely a question of -

(a) science of weak points and strong.
(b) dividing up their numbers.
(c) instituting signs and signals.
(d) direct and indirect maneuvers.

Q3. Sun Tzu said: To ensure that your army may withstand the brunt of the enemy's attack and remain unshaken - this is effected by

(a) science of weak points and strong.
(b) instituting signs and signals.
(c) dividing up their numbers.
(d) direct and indirect maneuvers.

Q4. Sun Tzu said: That the impact of your army may be like a grindstone dashed against an egg - this is effected by

(a) instituting signs and signals.
(b) direct and indirect maneuvers.
(c) science of weak points and strong.
(d) Heaven and Earth

Q5. Sun Tzu said: In all fighting, the ___(i)___ method may be used for joining battle, but ___(ii)___ methods will be needed in order to secure victory.

(i)	(ii)
(a) indirect	direct
(b) direct	direct
(c) indirect	indirect
(d) direct	**indirect**

Q6. Sun Tzu said: _____, efficiently applied, are inexhaustible as Heaven and Earth, unending as the flow of rivers and streams; like the sun and moon, they end but to begin anew; like the four seasons, they pass away to return once more.

(a) Direct tactics
(b) Indirect tactics
(c) Direct and indirect tactics
(d) Science of weak points

Q7. Sun Tzu said: In battle, there are not more than two methods of attack - the ___(i)___ and the ___(ii)___; yet these two in combination give rise to an ___(iii)___ series of maneuvers.

(i)	(ii)	(iii)
(a) direct	endless	indirect
(b) endless	indirect	direct
(c) direct	**indirect**	**endless**
(d) indirect	endless	direct

Q8. Sun Tzu said: The direct and the indirect methods of attack lead on to each other in turn. It is like moving in a circle - you never come to an end. (**True** or False)

Q9. Sun Tzu said: The ___(i)___ of troops is like the rush of a torrent which will even roll stones along in its course. The ___(ii)___ of decision is like the well-timed swoop of a falcon which enables it to strike and destroy its victim. Therefore the good fighter will be ___(iii)___ in his onset, and ___(iv)___ in his decision.

(i)	(ii)	(iii)	(iv)
(a) quality	onset	prompt	terrible
(b) quality	onset	terrible	prompt
(c) onset	**quality**	**terrible**	**prompt**
(d) onset	quality	prompt	terrible

Q10. Sun Tzu said: ___(i)___ may be likened to the ___(ii)___ of a crossbow; ___(iii)___, to the ___(iv)___ of a trigger.

(i)	(ii)	(iii)	(iv)
(a) decision	releasing	energy	bending
(b) decision	bending	energy	releasing
(c) energy	releasing	decision	bending
(d) energy	**bending**	**decision**	**releasing**

Q11. Sun Tzu said: Amid the turmoil and tumult of ___(i)___, there may be seeming disorder and yet no real ___(ii)___ at all; amid confusion and chaos, your array may be without ___(iii)___, yet it will be proof against ___(iv)___.

(i)	(ii)	(iii)	(iv)
(a) defeat	head or tail	disorder	battle
(b) disorder	defeat	head or tail	battle
(c) battle	**disorder**	**head or tail**	**defeat**
(d) head or tail	battle	defeat	disorder

Q12. Sun Tzu said: Hide order beneath the cloak of disorder. Conceal courage under a show of timidity. (**True** or False)

Q13. Sun Tzu said: One who is skillful at keeping the enemy on the move maintains ___(i)___ appearances, according to which the enemy will act. He ___(ii)___ something, that the enemy may snatch at it. By holding out baits, he keeps him on the ___(iii)___; then with a body of picked men he lies in ___(iv)___ for him.

(i)	(ii)	(iii)	(iv)
(a) march	wait	deceitful	sacrifices
(b) wait	march	sacrifices	deceitful
(c) deceitful	**sacrifices**	**march**	**wait**
(d) sacrifices	wait	march	deceitful

Q14. Sun Tzu said: The clever combatant looks to the effect of ___(i)___ energy, and does not require too much from ___(ii)___. Hence his ability to pick out the right ___(iii)___ and utilize combined ___(iv)___.

(i)	(ii)	(iii)	(iv)
(a) individuals	combined	energy	men
(b) men	individuals	combined	energy
(c) individuals	energy	combined	men
(d) combined	**individuals**	**men**	**energy**

Q15. Sun Tzu said: The ___(i)___ developed by good fighting men is as the ___(ii)___ of a round stone rolled down a mountain thousands of feet in height.

(i)	(ii)
(a) momentum	energy
(b) energy	energy
(c) momentum	momentum
(d) energy	**momentum**

6

WEAK POINTS AND STRONG

Q1. Sun Tzu said: Whoever is first in the field and ___(i)___ the coming of the enemy, will be ___(ii)___ for the fight; whoever is ___(iii)___ in the field and has to hasten to battle will arrive ___(iv)___.

(i)	(ii)	(iii)	(iv)
(a) second	exhausted	fresh	awaits
(b) awaits	exhausted	second	fresh
(c) exhausted	awaits	fresh	second
(d) awaits	**fresh**	**second**	**exhausted**

Q2. Sun Tzu said: The clever combatant imposes his will on the enemy, but does not allow the enemy's will to be imposed on him. (**True** or False)

Q3. What can a clever combatant do by holding out advantages to him?

(a) He can cause the enemy to approach of his own accord
(b) By inflicting damage, he can make it impossible for the enemy to draw near.
(c) If the enemy is taking his ease, he can harass him; if well supplied with food, he can starve him out.
(d) If enemy is quietly encamped, he can force enemy to move.

Q4. Sun Tzu said: Appear at points which the ___(i)___ must hasten to ___(ii)___; march swiftly to ___(iii)___ where you are ___(iv)___.

	(i)	(ii)	(iii)	(iv)
(a)	not expected	places	defend	enemy
(b)	enemy	places	defend	not expected
(c)	not expected	defend	enemy	places
(d)	**enemy**	**defend**	**places**	**not expected**

Q5. Sun Tzu said: An army may march great distances without distress, if it marches through country where the enemy is not. (**True** or False)

Q6. Sun Tzu said: You can be sure of succeeding in your ___(i)___ if you only attack places which are ___(ii)___. You can ensure the safety of your ___(iii)___ if you only hold positions that cannot be ___(iv)___.

	(i)	(ii)	(iii)	(iv)
(a)	defense	undefended	attacks	attacked
(b)	defense	attacked	attacks	undefended
(c)	**attacks**	**undefended**	**defense**	**attacked**
(d)	attacks	attacked	defense	undefended

Q7. Sun Tzu said: A general is skillful in attack whose opponent does not know what to ___(i)___; and he is skillful in ___(ii)___ whose opponent does not know what to ___(iii)___.

	(i)	(ii)	(iii)
(a)	attack	defense	defend
(b)	defend	attack	defense
(c)	attack	defend	defense
(d)	**defend**	**defense**	**attack**

Q8. Sun Tzu said: You may ___(i)___ and be absolutely irresistible, if you make for the enemy's ___(ii)___ points; you may retire and be ___(iii)___ from pursuit if your movements are more ___(iv)___ than those of the enemy.

	(i)	(ii)	(iii)	(iv)
(a)	weak	advance	rapid	safe

(b) weak	advance	safe	rapid
(c) advance	**weak**	**safe**	**rapid**
(d) advance	weak	rapid	safe

Q9. Sun Tzu said: If we wish to ___(i)___, the enemy can be ___(ii)___ to an engagement even though he be sheltered behind a high rampart and a deep ditch. All we need do is ___(iii)___ some other place that he will be obliged to ___(iv)___.

(i)	(ii)	(iii)	(iv)
(a) relieve	forced	fight	attack
(b) fight	**forced**	**attack**	**relieve**
(c) fight	relieve	attack	forced
(d) attack	forced	fight	relieve

Q10. Sun Tzu said: By discovering the enemy's ___(i)___ and remaining ___(ii)___ ourselves, we can keep our forces ___(iii)___, while the enemy's must be ___(iv)___.

(i)	(ii)	(iii)	(iv)
(a) invisible	concentrated	dispositions	divided
(b) invisible	concentrated	divided	dispositions
(c) dispositions	**invisible**	**concentrated**	**divided**
(d) dispositions	invisible	divided	concentrated

Q11. Sun Tzu said: We can form a single ___(i)___ body, while the enemy must ___(ii)___ up into fractions. Hence there will be a ___(iii)___ pitted against separate ___(iv)___ of a whole, which means that we shall be many to the enemy's few.

(i)	(ii)	(iii)	(iv)
(a) whole	parts	united	split
(b) parts	whole	united	split
(c) united	split	parts	whole
(d) united	**split**	**whole**	**parts**

Q12. Sun Tzu said: The spot where we intend to fight must not be made ___(i)___; for then the enemy will have to prepare against a possible attack at ___(ii)___ different points; and his forces being thus ___(iii)___ in many directions, the numbers we shall have to face at any given point will be proportionately ___(iv)___.

(i)	(ii)	(iii)	(iv)
(a) several	known	few	distributed
(b) several	known	distributed	few
(c) known	several	few	distributed
(d) known	**several**	**distributed**	**few**

Q13. Sun Tzu said: If enemy sends reinforcements everywhere, he will everywhere be strong. (True or **False**)

Q14. Sun Tzu said: Numerical ___(i)___ comes from having to prepare against possible attacks; numerical ___(ii)___, from compelling our adversary to make these preparations against us.

(i)	(ii)
(a) strength	weakness
(b) weakness	weakness
(c) strength	strength
(d) weakness	**strength**

Q15. Sun Tzu said: Knowing the place and the time of the coming battle, we may concentrate from the greatest distances in order to fight.
(**True** or False)

Q16 Sun Tzu said: If neither time nor place be known, then

(a) the left wing will be impotent to succor the right
(b) the right wing will be impotent to succor the left
(c) the van will be unable to relieve the rear
(d) the rear will be unable to support the van

Q17. Though the enemy be stronger in numbers, we may prevent him from fighting. How?

(a) Scheme so as to discover his plans and the likelihood of their success.
(b) Rouse him, and learn the principle of his activity or inactivity.
(c) Force him to reveal himself, so as to find out his vulnerable spots.
(d) Carefully compare the opposing army with your own, so that you may know where strength is superabundant and where it is deficient.

Q18. Sun Tzu said: In making tactical dispositions, the highest pitch you can attain is to ___(i)___ them; ___(ii)___ your dispositions, and you will be ___(iii)___ from the prying of the subtlest ___(iv)___, from the machinations of the wisest brains.

(i)	(ii)	(iii)	(iv)
(a) conceal	spies	conceal	safe
(b) conceal	spies	safe	conceal
(c) spies	conceal	conceal	safe
(d) conceal	**conceal**	**safe**	**spies**

Q19. Sun Tzu said: How ___(i)___ may be produced for them out of the ___(ii)___ own ___(iii)___ - that is what the ___(iv)___ cannot comprehend.

(i)	(ii)	(iii)	(iv)
(a) tactics	multitude	victory	enemy's
(b) tactics	multitude	enemy's	victory
(c) victory	**enemy's**	**tactics**	**multitude**
(d) victory	multitude	tactics	enemy's

Q20. Sun Tzu said: All men can see the ___(i)___ whereby I ___(ii)___, but what none can see is the ___(iii)___ out of which ___(iv)___ is evolved.

(i)	(ii)	(iii)	(iv)
(a) strategy	victory	tactics	conquer
(b) tactics	**conquer**	**strategy**	**victory**
(c) strategy	conquer	tactics	victory

(d) victory   conquer   tactics   strategy

Q21. Sun Tzu said: Do not repeat the tactics which have gained you one victory, but let your methods be regulated by the infinite variety of circumstances. (**True** or False)

Q22. Sun Tzu said: Military ___(i)___ are like unto ___(ii)___; for water in its natural course runs away from high places and hastens downwards. So in war, the way is to avoid what is ___(iii)___ and to strike at what is ___(iv)___ .

(i)	(ii)	(iii)	(iv)
(a) water	tactics	weak	strong
(b) weak	water	tactics	strong
(c) tactics	**water**	**strong**	**weak**
(d) tactics	water	weak	strong

Q23. Sun Tzu said: Water shapes its ___(i)___ according to the ___(ii)___ of the ground over which it flows; the soldier works out his ___(iii)___ in relation to the ___(iv)___ whom he is facing.

(i)	(ii)	(iii)	(iv)
(a) nature	course	foe	victory
(b) foe	nature	course	victory
(c) victory	course	foe	nature
(d) course	**nature**	**victory**	**foe**

Q24. Sun Tzu said: Just as water retains no constant ___(i)___ , so in warfare there are no constant ___(ii)___ . He who can ___(iii)___ his tactics in relation to his ___(iv)___ and thereby succeed in winning, may be called a heaven-born captain.

(i)	(ii)	(iii)	(iv)
(a) conditions	modify	shape	opponent
(b) conditions	opponent	shape	modify
(c) shape	**conditions**	**modify**	**opponent**
(d) shape	opponent	modify	conditions

7
MANEUVERING

Q1. Sun Tzu said: In war, the general having collected an army and concentrated his forces, he must ___(i)___ the different elements thereof before pitching his ___(ii)___. After that, comes ___(iii)___, than which there is nothing more difficult.

(i)	(ii)	(iii)
(a) camp	tactical maneuvering	blend and harmonize
(b) blend and harmonize	tactical maneuvering	camp
(c) camp	blend and harmonize	tactical maneuvering
(d) blend and harmonize	**camp**	**tactical maneuvering**

Q2. Sun Tzu said: The ___(i)___ of tactical maneuvering consists in turning the devious into the ___(ii)___, and misfortune into ___(iii)___.

(i)	(ii)	(iii)
(a) gain	direct	difficulty
(b) difficulty	**direct**	**gain**
(c) gain	direct	difficulty
(d) difficulty	gain	direct

Q3. Sun Tzu said: To take a long and circuitous route, after enticing the ___(i)___ out of the way, and though starting after him, to contrive to reach the ___(ii)___ before him, shows ___(iii)___ of the artifice of ___(iv)___.

	(i)	(ii)	(iii)	(iv)
(a)	goal	enemy	deviation	knowledge
(b)	goal	knowledge	deviation	enemy
(c)	**enemy**	**goal**	**knowledge**	**deviation**
(d)	enemy	knowledge	deviation	goal

Q4. Sun Tzu said: ___(i)___ with an army is ___(ii)___; with an ___(iii)___ multitude, most ___(iv)___.

	(i)	(ii)	(iii)	(iv)
(a)	undisciplined	dangerous	maneuvering	advantageous
(b)	undisciplined	advantageous	maneuvering	dangerous
(c)	maneuvering	dangerous	undisciplined	advantageous
(d)	**maneuvering**	**advantageous**	**undisciplined**	**dangerous**

Q5. Sun Tzu said: If you set a fully equipped ___(i)___ in march in order to snatch an ___(ii)___, the chances are that you will be too late. On the other hand, to detach a ___(iii)___ for the purpose involves the ___(iv)___ of its baggage and stores.

	(i)	(ii)	(iii)	(iv)
(a)	flying column	army	sacrifice	advantage
(b)	flying column	army	advantage	sacrifice
(c)	**army**	**advantage**	**flying column**	**sacrifice**
(d)	army	sacrifice	flying column	advantage

Q6. Sun Tzu said: We cannot enter into ___(i)___ until we are ___(ii)___ with the ___(iii)___ of our ___(iv)___.

	(i)	(ii)	(iii)	(iv)
(a)	designs	acquainted	neighbors	alliances
(b)	neighbors	alliances	designs	acquainted
(c)	designs	alliances	neighbors	acquainted
(d)	**alliances**	**acquainted**	**designs**	**neighbors**

Q7. Sun Tzu said: We are not fit to ___(i)___ an army on the ___(ii)___ unless we are familiar with the ___(iii)___ of the ___(iv)___ - its mountains and forests, its pitfalls and precipices, its marshes and swamps.

(i)	(ii)	(iii)	(iv)
(a) march	lead	country	face
(b) lead	**march**	**face**	**country**
(c) march	lead	face	country
(d) lead	country	march	face

Q8. Sun Tzu said: We shall be unable to turn natural advantage to account unless we make use of local guides. (**True** or False)

Q9. Sun Tzu said: In war, practice dissimulation, and you will not succeed. (True or **False**)

Q10. Sun Tzu said: Whether to concentrate or to divide your troops, must be decided by circumstances. (**True** or False)

Q11. Sun Tzu did not say:

(a) In raiding and plundering be like wind, in immovability like a mountain.
(b) Let your plans be dark and impenetrable as night.
(c) Let your rapidity be that of the wind, your compactness that of the forest.
(d) When you move, fall like a thunderbolt.

Q12. Sun Tzu said: When you plunder a countryside, let the spoil be ___(i)___ amongst your ___(ii)___; when you capture new territory, cut it up into allotments for the ___(iii)___ of the ___(iv)___.

(i)	(ii)	(iii)	(iv)
(a) benefit	men	soldiery	divided
(b) men	soldiery	benefit	divided
(c) divided	**men**	**benefit**	**soldiery**
(d) divided	benefit	men	soldiery

Q13. Sun Tzu said: Ponder and deliberate before you make a move. (**True** or False)

Q14. Sun Tzu said: He will ___(i)___ who has ___(ii)___ the artifice of ___(iii)___ . Such is the art of ___(iv)___ .

(i)	(ii)	(iii)	(iv)
(a) learnt	conquer	maneuvering	deviation
(b) learnt	conquer	deviation	maneuvering
(c) conquer	learnt	maneuvering	deviation
(d) conquer	**learnt**	**deviation**	**maneuvering**

Q15. Sun Tzu said: The Book of Army Management says: On the field of battle, the ___(i)___ does not carry far enough: hence the institution of ___(ii)___ . Nor can ordinary objects be ___(iii)___ clearly enough: hence the institution of ___(iv)___ .

(i)	(ii)	(iii)	(iv)
(a) gongs and drums	spoken word	seen	banners and flags
(b) banners and flags	spoken word	seen	gongs and drums
(c) spoken word	**gongs and drums**	**seen**	**banners and flags**
(d) spoken word	banners and flags	gongs and drums	seen

Q16. Sun Tzu said: The host (army) forming a ___(i)___ united body, it is impossible either for the brave to ___(ii)___ alone, or for the cowardly to ___(iii)___ alone. This is the art of handling ___(iv)___ masses of men.

(i)	(ii)	(iii)	(iv)
(a) large	retreat	advance	single
(b) single	**advance**	**retreat**	**large**
(c) single	retreat	advance	large
(d) large	advance	retreat	single

132

Q17. Sun Tzu said: In night-fighting make much use of ___(i)___, and in fighting by day, of ___(ii)___, as a means of influencing the ___(iii)___ of your army.

(i)	(ii)	(iii)
(a) flags and banners	ears and eyes	signal-fires and drums
(b) flags and banners	signal-fires and drums	ears and eyes
(c) signal-fires and drums	ears and eyes	flags and banners
(d) signal-fires and drums	**flags and banners**	**ears and eyes**

Q18. Sun Tzu said: A whole army may be robbed of its spirit; a commander-in-chief may be robbed of his presence of mind. (**True** or False)

Q19. Sun Tzu said: A soldier's spirit is ___(i)___ in the morning; by noonday it has begun to ___(ii)___; and in the evening, his mind is bent only on ___(iii)___ to camp.

(i)	(ii)	(iii)
(a) returning	flag	keenest
(b) returning	keenest	flag
(c) keenest	**flag**	**returning**
(d) flag	keenest	returning

Q20. Sun Tzu said: A clever general ___(i)___ an army when its spirit is ___(ii)___, but ___(iii)___ it when it is ___(iv)___ and inclined to return to camp. This is the art of studying moods.

(i)	(ii)	(iii)	(iv)
(a) attacks	keen	avoids	sluggish
(b) attacks	sluggish	avoids	keen
(c) avoids	sluggish	attacks	keen
(d) avoids	**keen**	**attacks**	**sluggish**

Q21. Sun Tzu said: Disciplined and calm, to ___(i)___ the appearance of ___(ii)___ and hubbub amongst the ___(iii)___:- this is the art of retaining ___(iv)___.

	(i)	(ii)	(iii)	(iv)
(a)	disorder	self-possession	await	enemy
(b)	**await**	**disorder**	**enemy**	**self-possession**
(c)	await	self-possession	disorder	enemy
(d)	disorder	self-possession	enemy	await

Q22. Sun Tzu said: To be ___(i)___ the goal while the enemy is still ___(ii)___ from it, to wait at ___(iii)___ while the enemy is toiling and ___(iv)___, to be well-fed while the enemy is famished :- this is the art of husbanding one's strength.

	(i)	(ii)	(iii)	(iv)
(a)	struggling	far	ease	near
(b)	near	struggling	far	ease
(c)	struggling	far	ease	near
(d)	**near**	**far**	**ease**	**struggling**

Q23. Which of the following is (are) true of the art of studying circumstances?

(a) Refrain from intercepting an enemy whose banners are in perfect order.
(b) Attack an army drawn up in calm and confident array.
(c) Intercept an enemy whose banners are in perfect order.
(d) Refrain from attacking an army drawn up in calm and confident array.

Q24. Sun Tzu said: It is a military axiom not to ___(i)___ against the enemy, nor to ___(ii)___ him when he comes ___(iii)___.

	(i)	(ii)	(iii)
(a)	oppose	downhill	advance uphill
(b)	**advance uphill**	**oppose**	**downhill**
(c)	advance uphill	downhill	advance uphill
(d)	downhill	oppose	advance uphill

Q25. Sun Tzu did not say:

(a) Do not pursue an enemy who simulates flight.
(b) Do attack soldiers whose temper is keen.
(c) Do not swallow bait offered by the enemy.
(d) Do not interfere with an army that is returning home.

Q26. Sun Tzu said: When you ___(i)___ an army, ___(ii)___ an outlet free. Do not ___(iii)___ a desperate foe too hard. Such is the art of warfare.

(i)	(ii)	(iii)
(a) leave	press	surround
(b) press	surround	leave
(c) surround	press	leave
(d) surround	**leave**	**press**

8
VARIATION IN TACTICS

Q1. Sun Tzu said: In war, the general receives his commands from the sovereign, collects his army and concentrates his forces. (**True** or False)

Q2. Sun Tzu said: When in difficult country, do not ___(i)___. In country where high roads intersect, join hands with your ___(ii)___. Do not linger in dangerously ___(iii)___ positions. In hemmed-in situations, you must resort to stratagem. In desperate position, you must ___(iv)___.

(i)	(ii)	(iii)	(iv)
(a) fight	isolated	allies	encamp
(b) fight	isolated	encamp	allies
(c) encamp	**allies**	**isolated**	**fight**
(d) encamp	isolated	allies	fight

Q3. Sun Tzu said: There are roads which must not be ___(i)___, armies which must be not ___(ii)___, towns which must be ___(iii)___, positions which must not be ___(iv)___, commands of the sovereign which must not be obeyed.

(i)	(ii)	(iii)	(iv)
(a) besieged	contested	attacked	followed
(b) followed	contest	attacked	besieged
(c) besieged	contested	followed	attacked
(d) followed	**attacked**	**besieged**	**contested**

136

Q4. Sun Tzu said: The general who thoroughly understands the ___(i)___ that accompany ___(ii)___ of tactics knows how to handle his troops. The general who does not understand these, may be well acquainted with the configuration of the ___(iii)___, yet he will not be able to turn his knowledge to ___(iv)___ account.

(i)	(ii)	(iii)	(iv)
(a) country	advantages	variation	practical
(b) variation	practical	advantages	country
(c) advantages	**variation**	**country**	**practical**
(d) advantages	practical	variation	country

Q5. Sun Tzu said: The student of war who is well versed in the art of war of varying his plans will fail to make the best use of his men. (True or **False**)

Q6. Sun Tzu said: In the wise leader's plans, considerations of advantage and of disadvantage will be blended together. (**True** or False)

Q7. Sun Tzu said: ___(i)___ the hostile chiefs by inflicting ___(ii)___ on them; and make ___(iii)___ for them, and keep them constantly ___(iv)___; hold out specious allurements, and make them rush to any given point.

(i)	(ii)	(iii)	(iv)
(a) trouble	reduce	damage	engaged
(b) reduce	**damage**	**trouble**	**engaged**
(c) reduce	trouble	damage	engaged
(d) engaged	trouble	reduce	damage

Q8. Sun Tzu said: The art of war teaches us to rely not on the likelihood of the enemy's ___(i)___, but on our own ___(ii)___ to receive him; not on the chance of his ___(iii)___, but rather on the fact that we have made our position ___(iv)___.

(i)	(ii)	(iii)	(iv)
(a) readiness	unassailable	not coming	not attacking
(b) unassailable	readiness	not attacking	not coming

(c) not coming   unassailable   not attacking   readiness

(d) not coming   readiness   not attacking   unassailable

Q9. Which of the following are besetting sin of a general, ruinous to the conduct of war ?

(a) Recklessness, which leads to destruction

(b) Cowardice, which leads to capture

(c) A hasty temper, which can be provoked by insults

(d) Over-solicitude for his men, which exposes him to worry and trouble.

Q10. Sun Tzu said: When an army is overthrown and its leader slain, the cause will surely be found among which dangerous faults?

(a) Recklessness

(b) Cowardice

(c) A hasty temper

(d) Over-solicitude for his men

9

THE ARMY ON THE MARCH

Q1. Sun Tzu said: ___(i)___ in high places, facing the sun. Do not climb ___(ii)___ in order to ___(iii)___. So much for ___(iv)___ warfare.

(i)	(ii)	(iii)	(iv)
(a) fight	mountain	camp	heights
(b) fight	heights	camp	mountain
(c) camp	mountain	fight	heights
(d) camp	**heights**	**fight**	**mountain**

Q2. Which of the following is (are) true of river warfare?

(a) When an invading force crosses a river in its onward march, do not advance to meet it in mid-stream. It will be best to let half the army get across, and then deliver your attack.
(b) After crossing a river, you should get far away from it.
(c) If you are anxious to fight, you should not go to meet the invader near a river which he has to cross.
(d) Moor your craft higher up than the enemy, and facing the sun. Do not move down-stream to meet the enemy.

Q3. Which of the following is (are) true of operations in salt-marshes?

(a) In crossing salt-marshes, your sole concern should be to get over them quickly, without any delay.
(b) Encamp at salt-marshes without any delay.

139

(c) If forced to fight in a salt-marsh, you should have water and grass near you, and get your back to a clump of trees.

(d) If forced to fight in a salt-marsh, you should get your back to a clump of trees.

Q4. Sun Tzu said: In campaigning in dry, level country, take up an easily accessible position with rising ground to your right and on your ___(i)___, so that the ___(ii)___ may be in ___(iii)___, and ___(iv)___ lie behind.

(i)	(ii)	(iii)	(iv)
(a) front	safety	rear	danger
(b) front	danger	rear	safety
(c) rear	**danger**	**front**	**safety**
(d) rear	safety	front	danger

Q5. What should be done in a country in which there are precipitous cliffs with torrents running between, deep natural hollows, confined places, tangled thickets, quagmires and crevasses?

(a) Keep away from such places.

(b) Get the enemy to approach such places.

(c) Leave such places with all possible speed.

(d) While such places are in our front, we should let the enemy have them on his rear.

Q6. Sun Tzu said: If in the ___(i)___ of your camp there should be any hilly country, ponds surrounded by aquatic grass, hollow basins filled with reeds, or woods with thick undergrowth, they must be carefully routed out and ___(ii)___; for these are places where men in ___(iii)___ or insidious ___(iv)___ are likely to be lurking.

(i)	(ii)	(iii)	(iv)
(a) ambush	searched	neighborhood	spies
(b) spies	ambush	neighborhood	searched
(c) neighborhood	**searched**	**ambush**	**spies**
(d) ambush	neighborhood	searched	spies

Q7. Sun Tzu said: When the enemy is ___(i)___ at hand and remains ___(ii)___, he is relying on the natural ___(iii)___ of his ___(iv)___.

	(i)	(ii)	(iii)	(iv)
(a)	quiet	close	position	strength
(b)	close	quiet	position	strength
(c)	quiet	close	strength	position
(d)	**close**	**quiet**	**strength**	**position**

Q8. Sun Tzu said: When enemy keeps ___(i)___ and tries to ___(ii)___ a battle, he is anxious for the ___(iii)___ to ___(iv)___.

	(i)	(ii)	(iii)	(iv)
(a)	other side	advance	aloof	provoke
(b)	**aloof**	**provoke**	**other side**	**advance**
(c)	other side	provoke	aloof	advance
(d)	aloof	advance	other side	provoke

Q9. Sun Tzu said: If place of encampment of enemy is easy of access, he is tendering a bait. (**True** or False)

Q10. Sun Tzu said: ___(i)___ amongst the trees of a forest shows that the enemy is ___(ii)___. The appearance of a number of screens in the midst of thick grass means that the ___(iii)___ wants to make us ___(iv)___.

	(i)	(ii)	(iii)	(iv)
(a)	enemy	advancing	suspicious	movement
(b)	suspicious	enemy	advancing	movement
(c)	**movement**	**advancing**	**enemy**	**suspicious**
(d)	advancing	movement	suspicious	enemy

Q11. Sun Tzu said: The ___(i)___ of birds in their flight is the sign of an ___(ii)___. ___(iii)___ beasts indicate that a sudden ___(iv)___ is coming.

	(i)	(ii)	(iii)	(iv)
(a)	attack	ambuscade	startled	rising

(b) ambuscade   startled   rising   attack

(c) attack   ambuscade   rising   startled

(d) rising   **ambuscade**   **startled**   **attack**

Q12. Sun Tzu said: When there is dust ___(i)___ in a high column, it is the sign of ___(ii)___ advancing; when the dust is ___(iii)___, but spread over a wide area, it betokens the approach of ___(iv)___.

(i)	(ii)	(iii)	(iv)
(a) low	infantry	rising	chariots
(b) rising	**chariots**	**low**	**infantry**
(c) low	chariots	rising	infantry
(d) rising	infantry	low	chariots

Q13. Sun Tzu said: When dust branches out in ___(i)___ directions, it shows that parties have been ___(ii)___ to collect firewood. A few clouds of dust moving ___(iii)___ signify that the army is ___(iv)___.

(i)	(ii)	(iii)	(iv)
(a) to and fro	encamping	different	sent
(b) different	encamping	to and fro	sent
(c) to and fro	sent	different	encamping
(d) different	**sent**	**to and fro**	**encamping**

Q14. Sun Tzu said: ___(i)___ words and increased preparations are signs that the enemy is about to ___(ii)___. ___(iii)___ language and driving forward as if to the attack are signs that he will ___(iv)___.

(i)	(ii)	(iii)	(iv)
(a) violent	advance	humble	retreat
(b) advance	retreat	violent	humble
(c) violent	retreat	humble	advance
(d) humble	**advance**	**violent**	**retreat**

Q15. Sun Tzu said: When the light chariots come out first and take up a position on the wings, it is a sign that the enemy is forming for battle. (**True** or False)

Q16. Sun Tzu said: ___(i)___ proposals ___(ii)___ by a ___(iii)___ covenant indicate a ___(iv)___.

(i)	(ii)	(iii)	(iv)
(a) unaccompanied	sworn	peace	plot
(b) sworn	plot	unaccompanied	peace
(c) peace	**unaccompanied**	**sworn**	**plot**
(d) unaccompanied	plot	sworn	peace

Q17. Sun Tzu said: When there is much running about and the soldiers fall out of rank, it means that the critical moment has come. (True or **False**)

Q18. Sun Tzu said: When some soldiers are seen advancing and some retreating, it is a lure. (**True** or False)

Q19. Sun Tzu said: If the soldiers who are sent to draw water begin by drinking themselves, the army is suffering from thirst. (**True** or False)

Q20. Sun Tzu said: If the enemy sees an advantage to be gained and makes no effort to secure it, the soldiers are exhausted. (**True** or False)

Q21. Sun Tzu said: If birds gather on any spot, it is unoccupied. (**True** or False)

Q22. Sun Tzu said: If there is disturbance in the camp, the general's authority is ___(i)___. If the banners and flags are shifted about, ___(ii)___ is afoot. If the officers are ___(iii)___, it means that the men are ___(iv)___.

(i)	(ii)	(iii)	(iv)
(a) weary	sedition	weak	angry
(b) angry	weak	sedition	weary
(c) weak	**sedition**	**angry**	**weary**

(d) sedition   weary   weak   angry

Q23. Sun Tzu said: The sight of men whispering together in small knots or speaking in subdued tones points to disaffection amongst the rank and file. (**True** or False)

Q24. Sun Tzu said: To begin by ___(i)___, but afterwards to take ___(ii)___ at the enemy's ___(iii)___, shows a supreme lack of ___(iv)___.

(i)	(ii)	(iii)	(iv)
(a) fright	bluster	intelligence	numbers
(b) bluster	numbers	intelligence	fright
(c) intelligence	fright	bluster	numbers
(d) bluster	**fright**	**numbers**	**intelligence**

Q25. Sun Tzu said: When ___(i)___ are sent with ___(ii)___ in their mouths, it is a sign that the ___(iii)___ wishes for a ___(iv)___.

(i)	(ii)	(iii)	(iv)
(a) truce	enemy	envoys	compliments
(b) enemy	truce	envoys	compliments
(c) envoys	**compliments**	**enemy**	**truce**
(d) compliments	truce	envoys	enemy

Q26. If our troops are no more in number than the enemy, that is amply sufficient, what can we do?

(a) Make no direct attack on enemy
(b) Concentrate all our available strength
(c) Keep a close watch on the enemy
(d) Obtain reinforcements

Q27. Sun Tzu said: He who exercises no ___(i)___ but makes ___(ii)___ of his ___(iii)___ is sure to be ___(iv)___ by them.

	(i)	(ii)	(iii)	(iv)
(a)	opponents	forethought	captured	light
(b)	**forethought**	**light**	**opponents**	**captured**
(c)	opponents	light	captured	forethought
(d)	light	opponents	captured	forethought

Q28. Sun Tzu said: If soldiers are ___(i)___ before they have grown ___(ii)___ to you, they will not prove ___(iii)___; and, unless submissive, then will be practically ___(iv)___.

	(i)	(ii)	(iii)	(iv)
(a)	useless	attached	submissive	punished
(b)	attached	submissive	useless	punished
(c)	submissive	useless	attached	punished
(d)	**punished**	**attached**	**submissive**	**useless**

Q29. Sun Tzu said: If, when the soldiers have become ___(i)___ to you, ___(ii)___ are not ___(iii)___, they will still be ___(iv)___.

	(i)	(ii)	(iii)	(iv)
(a)	useless	attached	punishments	enforced
(b)	punishments	useless	attached	enforced
(c)	**attached**	**punishments**	**enforced**	**useless**
(d)	enforced	punishments	useless	attached

Q30. Sun Tzu said: Soldiers must be treated in the first instance with ___(i)___, but kept under ___(ii)___ by means of iron ___(iii)___. This is a certain road to ___(iv)___.

	(i)	(ii)	(iii)	(iv)
(a)	victory	discipline	control	humanity
(b)	discipline	control	victory	humanity
(c)	**humanity**	**control**	**discipline**	**victory**
(d)	humanity	victory	control	discipline

Q31. Sun Tzu said: If in ___(i)___ soldiers ___(ii)___ are habitually ___(iii)___, the army will be well-disciplined; if not, its ___(iv)___ will be bad.

(i)	(ii)	(iii)	(iv)
(a) enforced	training	commands	discipline
(b) discipline	training	enforced	commands
(c) enforced	commands	discipline	training
(d) training	**commands**	**enforced**	**discipline**

10
TERRAIN

Q1. Sun Tzu said: We may distinguish six kinds of terrain, to wit: (1) Accessible ground; (2) ___(i)___; (3) temporizing ground; (4) ___(ii)___; (5) ___(iii)___; (6) positions at a great distance from the enemy.

(i)	(ii)	(iii)
(a) narrow passes	entangling ground	desperate ground
(b) entangling ground	**narrow passes**	**precipitous heights**
(c) precipitous heights	serious ground	narrow passes
(d) desperate ground	serious ground	entangling ground

Q2. Which of the following is (are) true of Accessible ground ?

(a) Be before the enemy in occupying the raised and sunny spots on accessible ground.
(b) Accessible ground can be freely traversed by both sides.
(c) With regard to ground of this nature, carefully guard your line of supplies.
(d) None of these.

Q3. Which of the following is (are) true of Entangling ground ?

(a) Entangling ground can be abandoned but is hard to re-occupy.
(b) If the enemy on entangling is unprepared, you may sally forth and defeat him.

(c) If the enemy on entangling is prepared for your coming, and you fail to defeat him, then, return being impossible, disaster will ensue.
(d) Only (a) and (c)

Q4. Sun Tzu said: When the position is such that neither side will gain by making _____ move, it is called temporizing ground.

(a) any
(b) no
(c) the first
(d) the second

Q5. Sun Tzu said: In a position of temporizing ground, even though the enemy should offer us an attractive ___(i)___, it will be advisable not to stir forth, but rather to ___(ii)___, thus ___(iii)___ the enemy in his turn; then, when part of his army has come out, we may deliver our ___(iv)___ with advantage.

(i)	(ii)	(iii)	(iv)
(a) enticing	attack	retreat	bait
(b) retreat	bait	enticing	attack
(c) attack	retreat	bait	enticing
(d) bait	**retreat**	**enticing**	**attack**

Q6. Sun Tzu said: With regard to narrow passes, if you can ___(i)___ them ___(ii)___, let them be strongly garrisoned and ___(iii)___ the advent of the ___(iv)___.

(i)	(ii)	(iii)	(iv)
(a) await	first	occupy	enemy
(b) await	enemy	occupy	first
(c) occupy	**first**	**await**	**enemy**
(d) occupy	enemy	await	first

Q7. Sun Tzu said: Should the army forestall you in occupying a pass, do not go after him if the pass is ___(i)___ garrisoned, but only if it is ___(ii)___ garrisoned.

	(i)	(ii)
(a)	fully	fully
(b)	weakly	fully
(c)	**fully**	**weakly**
(d)	weakly	weakly

Q8. Which of the following is (are) true of Precipitous heights ?

(a) If you reach precipitous heights before your adversary, you should occupy the raised and sunny spots, and there wait for him to come up.
(b) If the enemy has occupied precipitous heights before you, follow him and don't retreat.
(c) If the enemy has occupied them before you, do not follow him, but retreat and try to entice him away.
(d) None of these.

Q9. Sun Tzu said: If you are situated at a great ___(i)___ from the enemy, and the strength of the two armies is ___(ii)___, it is not easy to provoke a battle, and ___(iii)___ will be to your ___(iv)___ .

	(i)	(ii)	(iii)	(iv)
(a)	disadvantage	fighting	equal	distance
(b)	distance	fighting	equal	disadvantage
(c)	equal	distance	disadvantage	fighting
(d)	**distance**	**equal**	**fighting**	**disadvantage**

Q10. Sun Tzu said: There are six principles connected with Earth. The general who has attained a responsible post must be careful to study them. (**True** or False)

Q11. Sun Tzu said: An army is exposed to six several calamities, not arising from natural causes, but from faults for which the general is responsible.

These are: (1) Flight; (2) ___(i)___; (3) collapse; (4) ___(ii)___; (5) disorganization; (6) ___(iii)___.

(i)	(ii)	(iii)
(a) ruin	attack	rout
(b) insubordination	ruin	retreat
(c) rout	attack	insubordination
(d) insubordination	**ruin**	**rout**

Q12. Sun Tzu said: Other conditions being equal, if one force is hurled against another ten times its size, the result will be the flight of the former. (**True** or False)

Q13. Sun Tzu said: When the common soldiers are too ___(i)___ and their officers too ___(ii)___, the result is insubordination. When the officers are too ___(iii)___ and the common soldiers too ___(iv)___, the result is collapse.

(i)	(ii)	(iii)	(iv)
(a) weak	strong	strong	weak
(b) strong	**weak**	**strong**	**weak**
(c) strong	weak	weak	strong
(d) weak	strong	weak	strong

Q14. Sun Tzu said: When the higher officers are ___(i)___ and insubordinate, and on meeting the ___(ii)___ give battle on their own account from a feeling of ___(iii)___, before the commander-in-chief can tell whether or no he is in a position to ___(iv)___, the result is ruin.

(i)	(ii)	(iii)	(iv)
(a) enemy	angry	fight	resentment
(b) resentment	angry	enemy	fight
(c) angry	**enemy**	**resentment**	**fight**
(d) angry	fight	resentment	enemy

Q15. The result of which of the following is (are) utter disorganization?

(a) When the general is weak and without authority.
(b) When the general's orders are not clear and distinct.
(c) When there are no fixed duties assigned to officers and men.
(d) When the ranks are formed in a slovenly haphazard manner.

Q16. Sun Tzu said: When a general, unable to estimate the enemy's strength, allows an ___(i)___ force to engage a ___(ii)___ one, or hurls a ___(iii)___ detachment against a ___(iv)___ one, and neglects to place picked soldiers in the front rank, the result must be rout.

(i)	(ii)	(iii)	(iv)
(a) larger	inferior	powerful	weak
(b) powerful	weak	larger	inferior
(c) weak	inferior	powerful	larger
(d) inferior	**larger**	**weak**	**powerful**

Q17. Sun Tzu said: The natural formation of the country is the soldier's best ally; but a power of ___(i)___ the adversary, of ___(ii)___ the forces of victory, and of shrewdly calculating ___(iii)___, dangers and distances, constitutes the ___(iv)___ of a great general. He who knows these things, and in fighting puts his knowledge into practice, will win his battles. He who knows them not, nor practices them, will surely be defeated.

(i)	(ii)	(iii)	(iv)
(a) controlling	estimating	test	difficulties
(b) estimating	**controlling**	**difficulties**	**test**
(c) test	estimating	controlling	difficulties
(d) controlling	estimating	difficulties	test

Q18. Sun Tzu said: If fighting is sure to result in victory, then you ___(i)___ fight, even though the ruler ___(ii)___ it; if fighting will not result in victory, then you ___(iii)___ fight even at the ruler's ___(iv)___.

	(i)	(ii)	(iii)	(iv)
(a)	must not	forbid	must	bidding
(b)	must not	bidding	must	forbid
(c)	**must**	**forbid**	**must not**	**bidding**
(d)	forbid	must	must not	bidding

Q19. Sun Tzu said: The general who ___(i)___ without coveting ___(ii)___ and ___(iii)___ without fearing ___(iv)___, whose only thought is to protect his country and do good service for his sovereign, is the jewel of the kingdom.

	(i)	(ii)	(iii)	(iv)
(a)	retreats	fame	advances	disgrace
(b)	fame	disgrace	advances	retreats
(c)	retreats	disgrace	advances	fame
(d)	**advances**	**fame**	**retreats**	**disgrace**

Q20. Sun Tzu said: Regard your soldiers as your children, and they will follow you into the deepest valleys; look upon them as your own beloved sons, and they will stand by you even unto death. (**True** or False)

Q21. Sun Tzu said: If you are indulgent, but unable to make your ___(i)___ felt; kind-hearted, but unable to enforce your ___(ii)___; and incapable, moreover, of quelling ___(iii)___: then your soldiers must be likened to spoilt children; they are ___(iv)___ for any practical purpose.

	(i)	(ii)	(iii)	(iv)
(a)	disorder	authority	useless	commands
(b)	commands	authority	useless	disorder
(c)	useless	commands	disorder	authority
(d)	**authority**	**commands**	**disorder**	**useless**

Q22. Sun Tzu said: If we ___(i)___ that our own men are in a ___(ii)___ to attack, but are ___(iii)___ that the enemy is ___(iv)___ to attack, we have gone only halfway towards victory.

(i)	(ii)	(iii)	(iv)
(a) condition	not open	unaware	know
(b) unaware	not open	condition	know
(c) condition	unaware	know	not open
(d) know	**condition**	**unaware**	**not open**

Q23. Sun Tzu said: If we ___(i)___ that the enemy is ___(ii)___ to attack, but are ___(iii)___ that our own men are ___(iv)___ to attack, we have gone only halfway towards victory.

(i)	(ii)	(iii)	(iv)
(a) open	know	not in a condition	unaware
(b) know	**open**	**unaware**	**not in a condition**
(c) know	open	not in a condition	unaware
(d) open	know	unaware	not in a condition

Q24. Sun Tzu said: If we ___(i)___ that the enemy is ___(ii)___ to attack, and also know that our men are in a condition to attack, but are ___(iii)___ that the nature of the ___(iv)___ makes fighting impracticable, we have gone only halfway towards victory.

(i)	(ii)	(iii)	(iv)
(a) open	ground	know	unaware
(b) unaware	ground	open	know
(c) know	**open**	**unaware**	**ground**
(d) open	unaware	know	ground

Q25. Sun Tzu said: If you know the ___(i)___ and know ___(ii)___, your victory will not stand in doubt; if you know ___(iii)___ and know ___(iv)___, you may make your victory complete.

(i)	(ii)	(iii)	(iv)
(a) Heaven	yourself	enemy	Earth
(b) enemy	Earth	yourself	Heaven
(c) Earth	enemy	Heaven	yourself
(d) enemy	**yourself**	**Heaven**	**Earth**

11
THE NINE SITUATIONS

Q1. Sun Tzu said: The art of war recognizes nine varieties of ground: (1) Dispersive ground; (2) facile ground; (3) contentious ground; (4) ___(i)___; (5) ground of intersecting highways; (6) ___(ii)___; (7) difficult ground; (8) hemmed-in ground; (9) ___(iii)___ .

(i)	(ii)	(iii)
(a) open ground	entangling ground	serious ground
(b) serious ground	accessible ground	open ground
(c) open ground	**serious ground**	**desperate ground**
(d) desperate ground	open ground	entangling ground

Q2. Which of the following is (are) true of dispersive ground?

(a) When a chieftain is fighting in his own territory, it is dispersive ground.
(b) On dispersive ground, fight not.
(c) On dispersive ground, chieftain should inspire his men with unity of purpose.
(d) When chieftain has penetrated into hostile territory, but to no great distance, it is dispersive ground.

Q3. Which of the following is (are) true of facile ground?

(a) On facile ground, halt not.

154

(b) When chieftain has penetrated into hostile territory, but to no great distance, it is facile ground.

(c) Ground on which each side has liberty of movement is facile ground.

(d) On facile ground, chieftain would see that there is close connection between all parts of my army.

Q4. Which of the following is (are) true of contentious ground?

(a) On contentious ground, attack not.

(b) On contentious ground join hands with your allies.

(c) Ground the possession of which imports great advantage to either side, is contentious ground.

(d) On contentious ground, keep steadily on the march.

Q5. Which of the following is (are) NOT true of open ground?

(a) Ground on which each side has liberty of movement is open ground.

(b) On open ground, attack the enemy.

(c) On open ground, do not try to block the enemy's way.

(d) On open ground, keep a vigilant eye on your defenses.

Q6. Which of the following is (are) NOT true of ground of intersecting highways?

(a) Ground which forms the key to three contiguous states, so that he who occupies it first has most of the Empire at his command, is a ground of intersecting highways.

(b) On ground of intersecting highways, fight without delay.

(c) On ground of intersecting highways, join hands with allies.

(d) When there are means of communication on all four sides, the ground is one of intersecting highways.

Q7. Which of the following is (are) true of serious ground?

(a) On serious ground, try to ensure a continuous stream of supplies.

(b) On serious ground, gather in plunder.

(c) On serious ground don't try to block the enemy's way.

(d) When an army has penetrated into the heart of a hostile country, leaving a number of fortified cities in its rear, it is serious ground.

Q8. Which of the following is (are) NOT true of difficult ground?

(a) In difficult ground, keep steadily on the march.

(b) In difficult ground join hands with your allies.

(c) Mountain forests, rugged steeps, marshes and fens - all country that is hard to traverse: this is difficult ground.

(d) On difficult ground, don't encamp and keep pushing on along the road.

Q9. Which of the following is (are) true of hemmed-in ground?

(a) On hemmed-in ground, resort to stratagem.

(b) When you have the enemy's strongholds on your rear, and narrow passes in front, it is hemmed-in ground.

(c) When there is no place of refuge at all, it is hemmed-in ground.

(d) Ground which is reached through narrow gorges, and from which we can only retire by tortuous paths, so that a small number of the enemy would suffice to crush a large body of our men: this is hemmed in ground.

Q10. Which of the following is (are) NOT true of desperate ground?

(a) Ground on which we can only be saved from destruction by fighting without delay, is desperate ground.

(b) On desperate ground, fight not.

(c) On desperate ground proclaim to your soldiers the hopelessness of saving their lives.

(d) When there is no place of refuge at all, it is desperate ground.

Q11. Sun Tzu said: Those who were called skillful leaders of old knew how to -

(a) drive a wedge between the enemy's front and rear

(b) prevent co-operation between his large and small divisions

(c) hinder the good troops from rescuing the bad

(d) hinder the officers from rallying their men

Q12. Which of the following is (are) true of skillful leaders of old ?

(a) When the enemy's men were united, they managed to keep them in disorder.

(b) When it was to their advantage, they made a forward move.

(c) When the enemy's men were united, they managed to keep them in order.

(d) When it was to their disadvantage, they stopped still.

Q13. Sun Tzu said: If asked how to cope with a great host of the ___(i)___ in orderly array and on the point of marching to the ___(ii)___, I should say: "Begin by seizing something which your opponent holds ___(iii)___; then he will be ___(iv)___ to your will."

(i)	(ii)	(iii)	(iv)
(a) attack	enemy	amenable	dear
(b) enemy	dear	attack	dear
(c) attack	enemy	dear	amenable
(d) enemy	**attack**	**dear**	**amenable**

Q14. Sun Tzu said: ___(i)___ is the essence of war: take advantage of the enemy's ___(ii)___, make your way by ___(iii)___ routes, and attack ___(iv)___ spots.

(i)	(ii)	(iii)	(iv)
(a) unexpected	rapidity	unguarded	unreadiness
(b) unguarded	unreadiness	rapidity	unexpected
(c) rapidity	**unreadiness**	**unexpected**	**unguarded**
(d) unexpected	unreadiness	rapidity	unguarded

Q15. Which of the following is (are) the principles to be observed by an invading force?

(a) Make forays in fertile country in order to supply your army with food.

(b) The further you penetrate into a country, the greater will be the solidarity of your troops, and thus the defenders will not prevail against you.

(c) Carefully study the well-being of your men, and do not overtax them. Keep your army continually on the move, and devise unfathomable plans.

(d) Throw your soldiers into positions whence there is no escape, and they will prefer death to flight. If they will face death, there is nothing they may not achieve.

Q16. Sun Tzu said: Soldiers when in ___(i)___ straits lose the sense of ___(ii)___. If there is no place of refuge, they will stand firm. If they are in ___(iii)___ country, they will show a ___(iv)___ front. If there is no help for it, they will fight hard.

(i)	(ii)	(iii)	(iv)
(a) hostile	desperate	stubborn	fear
(b) desperate	**fear**	**hostile**	**stubborn**
(c) hostile	fear	desperate	stubborn
(d) desperate	stubborn	hostile	fear

Q17. Sun Tzu said: Prohibit the taking of ___(i)___, and do away with ___(ii)___ doubts. Then, until death itself comes, no ___(iii)___ need be ___(iv)___.

(i)	(ii)	(iii)	(iv)
(a) superstitious	feared	omens	calamity
(b) feared	superstitious	calamity	omens
(c) omens	**superstitious**	**calamity**	**feared**
(d) calamity	feared	omens	superstitious

Q18. Sun Tzu said: The principle on which to manage an army is to set up one standard of courage which all must reach. (**True** or False)

Q19. Which of the following is (are) true of the skillful general?

(a) The skillful general conducts his army just as though he were leading a single man, willy-nilly, by the hand.
(b) It is the business of a general to be quiet and thus ensure secrecy; upright and just, and thus maintain order.
(c) He must be able to mystify his officers and men by false reports and appearances, and thus keep them in total ignorance.
(d) By altering his arrangements and changing his plans, he keeps the enemy without definite knowledge. By shifting his camp and taking circuitous routes, he prevents the enemy from anticipating his purpose.

Q20. At the critical moment, how does the leader of an army acts ?

(a) He acts like one who has climbed up a height and then kicks away the ladder behind him.
(b) He carries his men deep into hostile territory before he shows his hand.
(c) He burns his boats and breaks his cooking-pots.
(d) Like a shepherd driving a flock of sheep, he drives his men this way and that, and nothing knows whither he is going.

Q21. Which of the following things that must most certainly be studied?

(a) The different measures suited to the nine varieties of ground.
(b) The expediency of aggressive or defensive tactics.
(c) The fundamental laws of human nature.
(d) None of these.

Q22. Sun Tzu said: When invading hostile territory, the general principle is, that penetrating ___(i)___ brings ___(ii)___; penetrating but a ___(iii)___ way means ___(iv)___.

(i)	(ii)	(iii)	(iv)
(a) short	cohesion	dispersion	deeply
(b) cohesion	dispersion	short	deeply

(c) dispersion   deeply   short   cohesion

(d) deeply   **cohesion**   **short**   **dispersion**

Q23. Sun Tzu said: It is the soldier's disposition to offer an obstinate ___(i)___ when ___(ii)___, to fight hard when he cannot help himself, and to ___(iii)___ promptly when he has fallen into ___(iv)___.

	(i)	(ii)	(iii)	(iv)
(a)	danger	resistance	surrounded	obey
(b)	obey	danger	surrounded	resistance
(c)	**resistance**	**surrounded**	**obey**	**danger**
(d)	danger	surrounded	obey	resistance

Q24. Sun Tzu said: We cannot enter into alliance with neighboring princes until we are acquainted with their designs. (**True** or False)

Q25. Sun Tzu said: We are fit to lead an army on the march unless we are unfamiliar with the face of the country - its mountains and forests, its pitfalls and precipices, its marshes and swamps. (True or **False**)

Q26. Sun Tzu said: We shall be unable to turn natural advantages to account unless we make use of local guides. (**True** or False)

Q27. Which of the following is (are) true of warlike prince ?

(a) When a warlike prince attacks a powerful state, his generalship shows itself in preventing the concentration of the enemy's forces.
(b) He overawes his opponents, and their allies are prevented from joining against him.
(c) He does not strive to ally himself with all and sundry, nor does he foster the power of other states.
(d) He carries out his own secret designs, keeping his antagonists in awe. Thus he is able to capture their cities and overthrow their kingdoms.

Q28. Sun Tzu said: Confront your soldiers with the ___(i)___ itself; never let them know your ___(ii)___. When the outlook is ___(ii)___, bring it before their eyes; but tell them nothing when the situation is ___(iv)___.

(i)	(ii)	(iii)	(iv)
(a) gloomy	bright	deed	design
(b) design	deed	gloomy	bright
(c) bright	design	gloomy	deed
(d) deed	**design**	**bright**	**gloomy**

Q29. Sun Tzu said: Place your army in deadly ___(i)___, and it will ___(ii)___; plunge it into desperate ___(iii)___, and it will come off in ___(iv)___. For it is precisely when a force has fallen into harm's way that is capable of striking a blow for victory.

(i)	(ii)	(iii)	(iv)
(a) straits	survive	safety	peril
(b) safety	peril	straits	survive
(c) straits	survive	peril	straits
(d) peril	**survive**	**straits**	**safety**

Q30. Sun Tzu said: Success in warfare is gained by carefully accommodating ourselves to the enemy's purpose. (**True** or False)

Q31. Sun Tzu said: Forestall your opponent by ___(i)___ what he holds ___(ii)___, and subtly contrive to ___(iii)___ his ___(iv)___ on the ground.

(i)	(ii)	(iii)	(iv)
(a) time	arrival	seizing	dear
(b) arrival	dear	time	seizing
(c) seizing	**dear**	**time**	**arrival**
(d) dear	seizing	time	arrival

Q32. Sun Tzu said: Walk in the path defined by rule, and accommodate yourself to the enemy until you can fight a decisive battle. (**True** or False)

Q33. Sun Tzu said: At first, then, exhibit the coyness of a maiden, until the ___(i)___ gives you an ___(ii)___; afterwards emulate the rapidity of a running hare, and it will be too ___(iii)___ for the enemy to ___(iv)___ you.

(i)	(ii)	(iii)	(iv)
(a) late	enemy	oppose	opening
(b) enemy	late	opening	oppose
(c) opening	enemy	oppose	late
(d) enemy	**opening**	**late**	**oppose**

12
THE ATTACK BY FIRE

Q1. Sun Tzu said: There are five ways of attacking with fire.
Which of the following is (are) part of those five ways?

(a) Burn soldiers in their camp
(b) Burn stores and burn baggage trains
(c) Burn arsenals and magazines
(d) Hurl dropping fire amongst the enemy

Q2. Sun Tzu said: In order to carry out an ___(i)___, we must have
___(ii)___ available. The material for raising ___(iii)___ should always be
kept in ___(iv)___.

	(i)	(ii)	(iii)	(iv)
(a)	fire	readiness	attack	means
(b)	fire	means	attack	readiness
(c)	**attack**	**means**	**fire**	**readiness**
(d)	attack	readiness	fire	means

Q3. Sun Tzu said: There is a proper season for making attacks with
___(i)___, and special days for starting a ___(ii)___. The proper season is
when the weather is ___(iii)___; the special days are those when the moon
is in the constellations of the Sieve, the Wall, the Wing or the Cross-bar; for
these four are all days of rising ___(iv)___.

	(i)	(ii)	(iii)	(iv)
(a)	wind	fire	conflagration	very dry
(b)	conflagration	wind	very dry	fire
(c)	**fire**	**conflagration**	**very dry**	**wind**
(d)	wind	fire	very dry	conflagration

Q4. Sun Tzu said: In attacking with fire, when fire breaks out ___(i)___ to enemy's ___(ii)___, respond at once with an ___(iii)___ from ___(iv)___.

	(i)	(ii)	(iii)	(iv)
(a)	camp	attack	inside	without
(b)	without	camp	attack	inside
(c)	attack	inside	without	camp
(d)	**inside**	**camp**	**attack**	**without**

Q5. Sun Tzu said: In attacking with fire, if there is an outbreak of fire, but the enemy's ___(i)___ remain ___(ii)___, bide your ___(iii)___ and do not ___(iv)___.

	(i)	(ii)	(iii)	(iv)
(a)	attack	quiet	soldiers	time
(b)	time	attack	soldiers	quiet
(c)	quiet	soldiers	attack	time
(d)	**soldiers**	**quiet**	**time**	**attack**

Q6. Sun Tzu said: In attacking with fire, when the force of the ___(i)___ has reached its ___(ii)___, follow it up with an ___(iii)___, if that is practicable; if not, ___(iv)___ where you are.

	(i)	(ii)	(iii)	(iv)
(a)	attack	height	flames	stay
(b)	height	stay	flames	attack
(c)	**flames**	**height**	**attack**	**stay**
(d)	stay	flames	height	attack

Q7. Sun Tzu said: In attacking with fire, if it is possible to make an ___(i)___ with fire from without, do not wait for it to break out ___(ii)___, but deliver your ___(iii)___ at a ___(iv)___ moment.

(i)	(ii)	(iii)	(iv)
(a) favorable	attack	assault	within
(b) attack	assault	within	favorable
(c) assault	**within**	**attack**	**favorable**
(d) attack	favorable	within	assault

Q8. Sun Tzu said: In attacking with fire, when you ___(i)___ a fire, be to ___(ii)___ of it. Do not ___(iii)___ from the ___(iv)___.

(i)	(ii)	(iii)	(iv)
(a) attack	leeward	start	windward
(b) start	**windward**	**attack**	**leeward**
(c) windward	leeward	attack	start
(d) leeward	windward	start	attack

Q9. Sun Tzu said: A wind that rises in the ___(i)___ lasts ___(ii)___, but a ___(iii)___ breeze soon ___(iv)___.

(i)	(ii)	(iii)	(iv)
(a) night	daytime	long	falls
(b) falls	long	night	daytime
(c) long	night	daytime	falls
(d) daytime	**long**	**night**	**falls**

Q10. Sun Tzu said: In every army, the five developments connected with ___(i)___ must be known, the movements of the ___(ii)___ calculated, and a ___(iii)___ kept for the proper ___(iv)___.

(i)	(ii)	(iii)	(iv)
(a) stars	days	fire	watch
(b) days	fire	watch	stars
(c) fire	**stars**	**watch**	**days**

(d) watch   days   fire   stars

Q11. Sun Tzu said: Those who use ___(i)___ as an aid to the attack show ___(ii)___; those who use ___(iii)___ as an aid to the attack gain an accession of ___(iv)___.

(i)	(ii)	(iii)	(iv)
(a) intelligence	strength	water	fire
(b) water	intelligence	strength	fire
(c) fire	**intelligence**	**water**	**strength**
(d) strength	fire	intelligence	water

Q12. Sun Tzu said: By means of water, an enemy may be intercepted, but not robbed of all his belongings. (**True** or False)

Q13. Sun Tzu said: Unhappy is the fate of one who tries to win his battles and succeed in his ___(i)___ without cultivating the spirit of ___(ii)___; for the result is waste of ___(iii)___ and general ___(iv)___.

(i)	(ii)	(iii)	(iv)
(a) enterprise	time	attacks	stagnation
(b) time	attacks	stagnation	enterprise
(c) stagnation	time	enterprise	attacks
(d) attacks	**enterprise**	**time**	**stagnation**

Q14. Sun Tzu said: The enlightened ___(i)___ lays his ___(ii)___ well ahead; the good general ___(iii)___ his ___(iv)___.

(i)	(ii)	(iii)	(iv)
(a) resources	cultivates	plans	ruler
(b) plans	ruler	resources	cultivates
(c) resources	plans	cultivates	ruler
(d) ruler	**plans**	**cultivates**	**resources**

Q15. Sun Tzu said: ___(i)___ not unless you see an ___(ii)___; use not your troops unless there is something to be gained; ___(iii)___ not unless the position is ___(iv)___ .

	(i)	(ii)	(iii)	(iv)
(a)	fight	critical	move	advantage
(b)	advantage	critical	move	fight
(c)	**move**	**advantage**	**fight**	**critical**
(d)	critical	fight	move	advantage

Q16. Sun Tzu said: No ___(i)___ should put troops into the field merely to gratify his own ___(ii)___; no ___(iii)___ should fight a battle simply out of ___(iv)___ .

	(i)	(ii)	(iii)	(iv)
(a)	general	ruler	spleen	pique
(b)	ruler	general	pique	spleen
(c)	**ruler**	**spleen**	**general**	**pique**
(d)	spleen	pique	ruler	general

Q17. Sun Tzu said: Anger may in time change to gladness; vexation may be succeeded by content. But a kingdom that has once been destroyed can never come again into being; nor can the dead ever be brought back to life. Hence the enlightened ruler is ___(i)___, and the good general full of ___(ii)___. This is the way to keep a country at ___(iii)___ and an army ___(iv)___ .

	(i)	(ii)	(iii)	(iv)
(a)	intact	peace	caution	heedful
(b)	peace	caution	intact	heedful
(c)	caution	heedful	intact	peace
(d)	**heedful**	**caution**	**peace**	**intact**

13

THE USE OF SPIES

Q1. Sun Tzu said: ___(i)___ a host of a hundred thousand men and ___(ii)___ them great distances entails heavy loss on the ___(iii)___ and a drain on the ___(iv)___ of the State.

	(i)	(ii)	(iii)	(iv)
(a)	marching	raising	resources	people
(b)	raising	people	marching	resources
(c)	marching	raising	resources	people
(d)	**raising**	**marching**	**people**	**resources**

Q2. Sun Tzu said: ___(i)___ armies may face each other for ___(ii)___, striving for the ___(iii)___ which is decided in a ___(iv)___ day.

	(i)	(ii)	(iii)	(iv)
(a)	single	victory	hostile	years
(b)	victory	hostile	years	single
(c)	**hostile**	**years**	**victory**	**single**
(d)	years	hostile	single	victory

Q3. Sun Tzu said: To remain in ___(i)___ of the enemy's ___(ii)___ simply because one grudges the outlay of a hundred ounces of silver in honors and emoluments, is the height of inhumanity. One who acts thus is no ___(iii)___ of men, no present help to his sovereign, no master of ___(iv)___.

	(i)	(ii)	(iii)	(iv)
(a)	condition	victory	ignorance	leader
(b)	victory	ignorance	leader	condition
(c)	**ignorance**	**condition**	**leader**	**victory**
(d)	leader	condition	victory	ignorance

Q4. Which of the following is (are) true of foreknowledge?

(a) Foreknowledge cannot be elicited from spirits.

(b) Foreknowledge cannot be obtained inductively from experience, nor by any deductive calculation.

(c) Foreknowledge enables the wise sovereign and the good general to strike and conquer, and achieve things beyond the reach of ordinary men.

(d) None of these.

Q5. Sun Tzu said: Knowledge of the enemy's dispositions can only be obtained from other men. Hence the use of spies. (**True** or False)

Q6. Sun Tzu said: The five classes of spies are : (1) Local spies; (2) ___(i)___ ; (3) converted spies; (4) ___(ii)___ ; (5) surviving spies.

	(i)	(ii)
(a)	outward spies	inverted spies
(b)	inward spies	outward spies
(c)	doomed spies	inverted spies
(d)	**inward spies**	**doomed spies**

Q7. Sun Tzu said: When the five kinds of ___(i)___ are all at ___(ii)___, none can discover the ___(iii)___ system. This is called "divine manipulation of the threads." It is the sovereign's most ___(iv)___ faculty.

	(i)	(ii)	(iii)	(iv)
(a)	secret	spy	precious	work
(b)	precious	secret	spy	work
(c)	**spy**	**work**	**secret**	**precious**
(d)	work	spy	precious	secret

Q8. Which if the following is (are) true of spies ?

(a) Having local spies means making use of officials of the enemy.
(b) Having inward spies means employing the services of the inhabitants of a district.
(c) Having converted spies, getting hold of the enemy's spies and using them for our own purposes.
(d) Surviving spies those who doing certain things openly for purposes of deception

Q9. Which if the following is (are) NOT true of spies ?

(a) Having local spies means making use of officials of the enemy.
(b) Surviving spies are those who bring back news from the enemy's camp.
(c) Having inward spies means employing the services of the inhabitants of a district.
(d) Doomed spies knowingly do certain things openly for purposes of deception, and report them to the enemy.

Q10. Sun Tzu said: Having ___(i)___ spies means employing the services of the inhabitants of a district. Having ___(ii)___ spies, making use of officials of the enemy. Having ___(iii)___ spies, getting hold of the enemy's spies and using them for our own purposes. ___(iv)___ spies, finally, are those who bring back news from the enemy's camp.

(i)	(ii)	(iii)	(iv)
(a) local	converted	inward	surviving
(b) inward	local	surviving	converted
(c) surviving	converted	local	inward
(d) local	**inward**	**converted**	**surviving**

Q11. Which of the following is (are) true of spies ?

(a) Spies cannot be usefully employed without a certain intuitive sagacity.

(b) None should be more liberally rewarded than the spies. None in the whole army are more intimate relations to be maintained than with spies.

(c) Spies can be properly managed without benevolence and straightforwardness.

(d) Without subtle ingenuity of mind, one cannot make certain of the truth of their reports.

Q12. Sun Tzu said: If a secret piece of news is ___(i)___ by a spy before the time is ___(ii)___, he must be put to ___(iii)___ together with the man to whom the secret was ___(iv)___.

(i)	(ii)	(iii)	(iv)
(a) told	divulged	ripe	death
(b) ripe	told	death	divulged
(c) divulged	**ripe**	**death**	**told**
(d) told	death	ripe	divulged

Q13. Sun Tzu said: Whether the object be to crush an army, to storm a city, or to assassinate an individual, it is always necessary to begin by ___(i)___ out the ___(ii)___ of the attendants, the aides-de-camp, and door-keepers and sentries of the general in command. Our ___(iii)___ must be commissioned to ___(iv)___ these.

(i)	(ii)	(iii)	(iv)
(a) ascertain	spies	names	finding
(b) finding	spies	names	ascertain
(c) ascertain	finding	spies	names
(d) finding	**names**	**spies**	**ascertain**

Q14. Sun Tzu said: The enemy's spies who have come to spy on us must be sought out, tempted with bribes, led away and comfortably housed. Thus they will become _____ spies and available for our service.

(a) inward

(b) converted

(c) doomed

(d) surviving

Q15. Sun Tzu said: It is essential that the converted spy be treated with the utmost liberality. Why ?

(a) It is through the information brought by the converted spy that we are able to acquire and employ local and inward spies.
(b) It is owing to the information brought by the converted spy, that we can cause the doomed spy to carry false tidings to the enemy.
(c) It is by information brought by the converted spy that the surviving spy can be used on appointed occasions.
(d) The end and aim of spying in all its five varieties is knowledge of the enemy; and this knowledge can only be derived, in the first instance, from the converted spy.

Q16. Sun Tzu said: It is only the enlightened ruler and the wise ___(i)___ who will use the highest ___(ii)___ of the army for purposes of ___(iii)___ and thereby they achieve great ___(iv)___ .

(i)	(ii)	(iii)	(iv)
(a) spying	results	intelligence	general
(b) intelligence	spying	general	results
(c) general	**intelligence**	**spying**	**results**
(d) results	general	intelligence	spying

Q17. Sun Tzu said: ___(i)___ are a most important element in ___(ii)___, because on them depends an ___(iii)___ ability to ___(iv)___ .

(i)	(ii)	(iii)	(iv)
(a) army's	war	spies	move
(b) war	spies	army's	move
(c) army's	move	spies	war
(d) spies	**war**	**army's**	**move**

PART III

This part contains *The Art of War* authored by Sun Tzu and translated into English by Lionel Giles.

The Art of War

By Sun Tzu

Translated By Lionel Giles

1
LAYING PLANS

1. Sun Tzu said: The art of war is of vital importance to the State.

2. It is a matter of life and death, a road either to safety or to ruin. Hence it is a subject of inquiry which can on no account be neglected.

3. The art of war, then, is governed by five constant factors, to be taken into account in one's deliberations, when seeking to determine the conditions obtaining in the field.

4. These are: (1) The Moral Law; (2) Heaven; (3) Earth; (4) The Commander; (5) Method and Discipline.

5. The Moral Law causes the people to be in complete accord with their ruler, so that they will follow him regardless of their lives, undismayed by any danger.

6. Heaven signifies night and day, cold and heat, times and seasons.

7. Earth comprises distances, great and small; danger and security; open ground and narrow passes; the chances of life and death.

8. The Commander stands for the virtues of wisdom, sincerely, benevolence, courage and strictness.

9. By Method and Discipline are to be understood the marshaling of the army in its proper subdivisions, the graduations of rank among the officers, the maintenance of roads by which supplies may reach the army, and the control of military expenditure.

10. These five heads should be familiar to every general: he who knows them will be victorious; he who knows them not will fail.

11. Therefore, in your deliberations, when seeking to determine the military conditions, let them be made the basis of a comparison, in this wise:-

12. (1) Which of the two sovereigns is imbued with the Moral law? (2) Which of the two generals has most ability? (3) With whom lie the advantages derived from Heaven and Earth? (4) On which side is discipline most rigorously enforced? (5) Which army is stronger? (6) On which side are officers and men more highly trained? (7) In which army is there the greater constancy both in reward and punishment?

13. By means of these seven considerations I can forecast victory or defeat.

14. The general that hearkens to my counsel and acts upon it, will conquer: let such a one be retained in command! The general that hearkens not to my counsel nor acts upon it, will suffer defeat:- let such a one be dismissed!

15. While heading the profit of my counsel, avail yourself also of any helpful circumstances over and beyond the ordinary rules.

16. According as circumstances are favorable, one should modify one's plans.

17. All warfare is based on deception.

18. Hence, when able to attack, we must seem unable; when using our forces, we must seem inactive; when we are near, we must make the enemy believe we are far away; when far away, we must make him believe we are near.

19. Hold out baits to entice the enemy. Feign disorder, and crush him.

20. If he is secure at all points, be prepared for him. If he is in superior strength, evade him.

21. If your opponent is of choleric temper, seek to irritate him. Pretend to be weak, that he may grow arrogant.

22. If he is taking his ease, give him no rest. If his forces are united, separate them.

23. Attack him where he is unprepared, appear where you are not expected.

24. These military devices, leading to victory, must not be divulged beforehand.

25. Now the general who wins a battle makes many calculations in his temple ere the battle is fought. The general who loses a battle makes but few calculations beforehand. Thus do many calculations lead to victory, and few calculations to defeat: how much more no calculation at all! It is by attention to this point that I can foresee who is likely to win or lose.

2

WAGING WAR

1. Sun Tzu said: In the operations of war, where there are in the field a thousand swift chariots, as many heavy chariots, and a hundred thousand mail-clad soldiers, with provisions enough to carry them a thousand li, the expenditure at home and at the front, including entertainment of guests, small items such as glue and paint, and sums spent on chariots and armor, will reach the total of a thousand ounces of silver per day. Such is the cost of raising an army of 100,000 men.

2. When you engage in actual fighting, if victory is long in coming, then men's weapons will grow dull and their ardor will be damped. If you lay siege to a town, you will exhaust your strength.

3. Again, if the campaign is protracted, the resources of the State will not be equal to the strain.

4. Now, when your weapons are dulled, your ardor damped, your strength exhausted and your treasure spent, other chieftains will spring up to take advantage of your extremity. Then no man, however wise, will be able to avert the consequences that must ensue.

5. Thus, though we have heard of stupid haste in war, cleverness has never been seen associated with long delays.

6. There is no instance of a country having benefited from prolonged warfare.

7. It is only one who is thoroughly acquainted with the evils of war that can thoroughly understand the profitable way of carrying it on.

8. The skillful soldier does not raise a second levy, neither are his supply-wagons loaded more than twice.

9. Bring war material with you from home, but forage on the enemy. Thus the army will have food enough for its needs.

10. Poverty of the State exchequer causes an army to be maintained by contributions from a distance. Contributing to maintain an army at a distance causes the people to be impoverished.

11. On the other hand, the proximity of an army causes prices to go up; and high prices cause the people's substance to be drained away.

12. When their substance is drained away, the peasantry will be afflicted by heavy exactions.

13. With this loss of substance and exhaustion of strength, the homes of the people will be stripped bare, and three-tenths of their income will be dissipated; while government expenses for broken chariots, worn-out horses, breast-plates and helmets, bows and arrows, spears and shields, protective mantles, draught-oxen and heavy wagons, will amount to four-tenths of its total revenue.

14. Hence a wise general makes a point of foraging on the enemy. One cartload of the enemy's provisions is equivalent to twenty of one's own, and likewise a single picul of his provender is equivalent to twenty from one's own store.

15. Now in order to kill the enemy, our men must be roused to anger; that there may be advantage from defeating the enemy, they must have their rewards.

16. Therefore in chariot fighting, when ten or more chariots have been taken, those should be rewarded who took the first. Our own flags should be substituted for those of the enemy, and the chariots mingled and used in conjunction with ours. The captured soldiers should be kindly treated and kept.

17. This is called, using the conquered foe to augment one's own strength.

18. In war, then, let your great object be victory, not lengthy campaigns.

19. Thus it may be known that the leader of armies is the arbiter of the people's fate, the man on whom it depends whether the nation shall be in peace or in peril.

3
ATTACK BY STRATAGEM

1. Sun Tzu said: In the practical art of war, the best thing of all is to take the enemy's country whole and intact; to shatter and destroy it is not so good. So, too, it is better to recapture an army entire than to destroy it, to capture a regiment, a detachment or a company entire than to destroy them.

2. Hence to fight and conquer in all your battles is not supreme excellence; supreme excellence consists in breaking the enemy's resistance without fighting.

3. Thus the highest form of generalship is to balk the enemy's plans; the next best is to prevent the junction of the enemy's forces; the next in order is to attack the enemy's army in the field; and the worst policy of all is to besiege walled cities.

4. The rule is, not to besiege walled cities if it can possibly be avoided. The preparation of mantlets, movable shelters, and various implements of war, will take up three whole months; and the piling up of mounds over against the walls will take three months more.

5. The general, unable to control his irritation, will launch his men to the assault like swarming ants, with the result that one-third of his men are slain, while the town still remains untaken. Such are the disastrous effects of a siege.

6. Therefore the skillful leader subdues the enemy's troops without any fighting; he captures their cities without laying siege to them; he overthrows their kingdom without lengthy operations in the field.

7. With his forces intact he will dispute the mastery of the Empire, and thus, without losing a man, his triumph will be complete. This is the method of attacking by stratagem.

8. It is the rule in war, if our forces are ten to the enemy's one, to surround him; if five to one, to attack him; if twice as numerous, to divide our army into two.

9. If equally matched, we can offer battle; if slightly inferior in numbers, we can avoid the enemy; if quite unequal in every way, we can flee from him.

10. Hence, though an obstinate fight may be made by a small force, in the end it must be captured by the larger force.

11. Now the general is the bulwark of the State; if the bulwark is complete at all points; the State will be strong; if the bulwark is defective, the State will be weak.

12. There are three ways in which a ruler can bring misfortune upon his army:-

13. (1) By commanding the army to advance or to retreat, being ignorant of the fact that it cannot obey. This is called hobbling the army.

14. (2) By attempting to govern an army in the same way as he administers a kingdom, being ignorant of the conditions which obtain in an army. This causes restlessness in the soldier's minds.

15. (3) By employing the officers of his army without discrimination, through ignorance of the military principle of adaptation to circumstances. This shakes the confidence of the soldiers.

16. But when the army is restless and distrustful, trouble is sure to come from the other feudal princes. This is simply bringing anarchy into the army, and flinging victory away.

17. Thus we may know that there are five essentials for victory: (1) He will win who knows when to fight and when not to fight. (2) He will win who knows how to handle both superior and inferior forces. (3) He will win whose army is animated by the same spirit throughout all its ranks. (4) He will win who, prepared himself, waits to take the enemy unprepared. (5) He will win who has military capacity and is not interfered with by the sovereign.

18. Hence the saying: If you know the enemy and know yourself, you need not fear the result of a hundred battles. If you know yourself but not the enemy, for every victory gained you will also suffer a defeat. If you know neither the enemy nor yourself, you will succumb in every battle.

4

TACTICAL DISPOSITIONS

1. Sun Tzu said: The good fighters of old first put themselves beyond the possibility of defeat, and then waited for an opportunity of defeating the enemy.

2. To secure ourselves against defeat lies in our own hands, but the opportunity of defeating the enemy is provided by the enemy himself.

3. Thus the good fighter is able to secure himself against defeat, but cannot make certain of defeating the enemy.

4. Hence the saying: One may know how to conquer without being able to do it.

5. Security against defeat implies defensive tactics; ability to defeat the enemy means taking the offensive.

6. Standing on the defensive indicates insufficient strength; attacking, a superabundance of strength.

7. The general who is skilled in defense hides in the most secret recesses of the earth; he who is skilled in attack flashes forth from the topmost heights of heaven. Thus on the one hand we have ability to protect ourselves; on the other, a victory that is complete.

8. To see victory only when it is within the ken of the common herd is not the acme of excellence.

9. Neither is it the acme of excellence if you fight and conquer and the whole Empire says, "Well done!"

10. To lift an autumn hair is no sign of great strength; to see the sun and moon is no sign of sharp sight; to hear the noise of thunder is no sign of a quick ear.

11. What the ancients called a clever fighter is one who not only wins, but excels in winning with ease.

12. Hence his victories bring him neither reputation for wisdom nor credit for courage.

13. He wins his battles by making no mistakes. Making no mistakes is what establishes the certainty of victory, for it means conquering an enemy that is already defeated.

14. Hence the skillful fighter puts himself into a position which makes defeat impossible, and does not miss the moment for defeating the enemy.

15. Thus it is that in war the victorious strategist only seeks battle after the victory has been won, whereas he who is destined to defeat first fights and afterwards looks for victory.

16. The consummate leader cultivates the moral law, and strictly adheres to method and discipline; thus it is in his power to control success.

17. In respect of military method, we have, firstly, Measurement; secondly, Estimation of quantity; thirdly, Calculation; fourthly, Balancing of chances; fifthly, Victory.

18. Measurement owes its existence to Earth; Estimation of quantity to Measurement; Calculation to Estimation of quantity; Balancing of chances to Calculation; and Victory to Balancing of chances.

19. A victorious army opposed to a routed one, is as a pound's weight placed in the scale against a single grain.

20. The onrush of a conquering force is like the bursting of pent-up waters into a chasm a thousand fathoms deep.

5
ENERGY

1. Sun Tzu said: The control of a large force is the same principle as the control of a few men: it is merely a question of dividing up their numbers.

2. Fighting with a large army under your command is nowise different from fighting with a small one: it is merely a question of instituting signs and signals.

3. To ensure that your whole host may withstand the brunt of the enemy's attack and remain unshaken - this is effected by maneuvers direct and indirect.

4. That the impact of your army may be like a grindstone dashed against an egg - this is effected by the science of weak points and strong.

5. In all fighting, the direct method may be used for joining battle, but indirect methods will be needed in order to secure victory.

6. Indirect tactics, efficiently applied, are inexhaustible as Heaven and Earth, unending as the flow of rivers and streams; like the sun and moon, they end but to begin anew; like the four seasons, they pass away to return once more.

7. There are not more than five musical notes, yet the combinations of these five give rise to more melodies than can ever be heard.

8. There are not more than five primary colors (blue, yellow, red, white, and black), yet in combination they produce more hues than can ever been seen.

9. There are not more than five cardinal tastes (sour, acrid, salt, sweet, bitter), yet combinations of them yield more flavors than can ever be tasted.

10. In battle, there are not more than two methods of attack--the direct and the indirect; yet these two in combination give rise to an endless series of maneuvers.

11. The direct and the indirect lead on to each other in turn. It is like moving in a circle - you never come to an end. Who can exhaust the possibilities of their combination?

12. The onset of troops is like the rush of a torrent which will even roll stones along in its course.

13. The quality of decision is like the well-timed swoop of a falcon which enables it to strike and destroy its victim.

14. Therefore the good fighter will be terrible in his onset, and prompt in his decision.

15. Energy may be likened to the bending of a crossbow; decision, to the releasing of a trigger.

16. Amid the turmoil and tumult of battle, there may be seeming disorder and yet no real disorder at all; amid confusion and chaos, your array may be without head or tail, yet it will be proof against defeat.

17. Simulated disorder postulates perfect discipline, simulated fear postulates courage; simulated weakness postulates strength.

18. Hiding order beneath the cloak of disorder is simply a question of subdivision; concealing courage under a show of timidity presupposes a fund

of latent energy; masking strength with weakness is to be effected by tactical dispositions.

19. Thus one who is skillful at keeping the enemy on the move maintains deceitful appearances, according to which the enemy will act. He sacrifices something, that the enemy may snatch at it.
20. By holding out baits, he keeps him on the march; then with a body of picked men he lies in wait for him.

21. The clever combatant looks to the effect of combined energy, and does not require too much from individuals. Hence his ability to pick out the right men and utilize combined energy.

22. When he utilizes combined energy, his fighting men become as it were like unto rolling logs or stones. For it is the nature of a log or stone to remain motionless on level ground, and to move when on a slope; if four-cornered, to come to a standstill, but if round-shaped, to go rolling down.

23. Thus the energy developed by good fighting men is as the momentum of a round stone rolled down a mountain thousands of feet in height. So much on the subject of energy.

6
WEAK POINTS AND STRONG

1. Sun Tzu said: Whoever is first in the field and awaits the coming of the enemy, will be fresh for the fight; whoever is second in the field and has to hasten to battle will arrive exhausted.

2. Therefore the clever combatant imposes his will on the enemy, but does not allow the enemy's will to be imposed on him.

3. By holding out advantages to him, he can cause the enemy to approach of his own accord; or, by inflicting damage, he can make it impossible for the enemy to draw near.

4. If the enemy is taking his ease, he can harass him; if well supplied with food, he can starve him out; if quietly encamped, he can force him to move.

5. Appear at points which the enemy must hasten to defend; march swiftly to places where you are not expected.

6. An army may march great distances without distress, if it marches through country where the enemy is not.

7. You can be sure of succeeding in your attacks if you only attack places which are undefended. You can ensure the safety of your defense if you only hold positions that cannot be attacked.

8. Hence that general is skillful in attack whose opponent does not know what to defend; and he is skillful in defense whose opponent does not know what to attack.

9. O divine art of subtlety and secrecy! Through you we learn to be invisible, through you inaudible; and hence we can hold the enemy's fate in our hands.

10. You may advance and be absolutely irresistible, if you make for the enemy's weak points; you may retire and be safe from pursuit if your movements are more rapid than those of the enemy.

11. If we wish to fight, the enemy can be forced to an engagement even though he be sheltered behind a high rampart and a deep ditch. All we need do is attack some other place that he will be obliged to relieve.

12. If we do not wish to fight, we can prevent the enemy from engaging us even though the lines of our encampment be merely traced out on the ground. All we need do is to throw something odd and unaccountable in his way.

13. By discovering the enemy's dispositions and remaining invisible ourselves, we can keep our forces concentrated, while the enemy's must be divided.

14. We can form a single united body, while the enemy must split up into fractions. Hence there will be a whole pitted against separate parts of a whole, which means that we shall be many to the enemy's few.

15. And if we are able thus to attack an inferior force with a superior one, our opponents will be in dire straits.

16. The spot where we intend to fight must not be made known; for then the enemy will have to prepare against a possible attack at several different points; and his forces being thus distributed in many directions, the numbers we shall have to face at any given point will be proportionately few.

17. For should the enemy strengthen his van, he will weaken his rear; should he strengthen his rear, he will weaken his van; should he strengthen his left, he will weaken his right; should he strengthen his right, he will weaken his left. If he sends reinforcements everywhere, he will everywhere be weak.

18. Numerical weakness comes from having to prepare against possible attacks; numerical strength, from compelling our adversary to make these preparations against us.

19. Knowing the place and the time of the coming battle, we may concentrate from the greatest distances in order to fight.

20. But if neither time nor place be known, then the left wing will be impotent to succor the right, the right equally impotent to succor the left, the van unable to relieve the rear, or the rear to support the van. How much more so if the furthest portions of the army are anything under a hundred LI apart, and even the nearest are separated by several LI!

21. Though according to my estimate the soldiers of Yueh exceed our own in number, that shall advantage them nothing in the matter of victory. I say then that victory can be achieved.

22. Though the enemy be stronger in numbers, we may prevent him from fighting. Scheme so as to discover his plans and the likelihood of their success.

23. Rouse him, and learn the principle of his activity or inactivity. Force him to reveal himself, so as to find out his vulnerable spots.

24. Carefully compare the opposing army with your own, so that you may know where strength is superabundant and where it is deficient.

25. In making tactical dispositions, the highest pitch you can attain is to conceal them; conceal your dispositions, and you will be safe from the prying of the subtlest spies, from the machinations of the wisest brains.

26. How victory may be produced for them out of the enemy's own tactics-- that is what the multitude cannot comprehend.

27. All men can see the tactics whereby I conquer, but what none can see is the strategy out of which victory is evolved.

28. Do not repeat the tactics which have gained you one victory, but let your methods be regulated by the infinite variety of circumstances.

29. Military tactics are like unto water; for water in its natural course runs away from high places and hastens downwards.

30. So in war, the way is to avoid what is strong and to strike at what is weak.

31. Water shapes its course according to the nature of the ground over which it flows; the soldier works out his victory in relation to the foe whom he is facing.

32. Therefore, just as water retains no constant shape, so in warfare there are no constant conditions.

33. He who can modify his tactics in relation to his opponent and thereby succeed in winning, may be called a heaven-born captain.

34. The five elements (water, fire, wood, metal, earth) are not always equally predominant; the four seasons make way for each other in turn. There are short days and long; the moon has its periods of waning and waxing.

7

MANEUVERING

1. Sun Tzu said: In war, the general receives his commands from the sovereign.

2. Having collected an army and concentrated his forces, he must blend and harmonize the different elements thereof before pitching his camp.

3. After that, comes tactical maneuvering, than which there is nothing more difficult. The difficulty of tactical maneuvering consists in turning the devious into the direct, and misfortune into gain.

4. Thus, to take a long and circuitous route, after enticing the enemy out of the way, and though starting after him, to contrive to reach the goal before him, shows knowledge of the artifice of deviation.

5. Maneuvering with an army is advantageous; with an undisciplined multitude, most dangerous.

6. If you set a fully equipped army in march in order to snatch an advantage, the chances are that you will be too late. On the other hand, to detach a flying column for the purpose involves the sacrifice of its baggage and stores.

7. Thus, if you order your men to roll up their buff-coats, and make forced marches without halting day or night, covering double the usual distance at

a stretch, doing a hundred LI in order to wrest an advantage, the leaders of all your three divisions will fall into the hands of the enemy.

8. The stronger men will be in front, the jaded ones will fall behind, and on this plan only one-tenth of your army will reach its destination.

9. If you march fifty LI in order to outmaneuver the enemy, you will lose the leader of your first division, and only half your force will reach the goal.

10. If you march thirty LI with the same object, two-thirds of your army will arrive.

11. We may take it then that an army without its baggage-train is lost; without provisions it is lost; without bases of supply it is lost.

12. We cannot enter into alliances until we are acquainted with the designs of our neighbors.

13. We are not fit to lead an army on the march unless we are familiar with the face of the country - its mountains and forests, its pitfalls and precipices, its marshes and swamps.

14. We shall be unable to turn natural advantage to account unless we make use of local guides.

15. In war, practice dissimulation, and you will succeed.

16. Whether to concentrate or to divide your troops, must be decided by circumstances.

17. Let your rapidity be that of the wind, your compactness that of the forest.

18. In raiding and plundering be like fire, is immovability like a mountain.

19. Let your plans be dark and impenetrable as night, and when you move, fall like a thunderbolt.

20. When you plunder a countryside, let the spoil be divided amongst your men; when you capture new territory, cut it up into allotments for the benefit of the soldiery.

21. Ponder and deliberate before you make a move.

22. He will conquer who has learnt the artifice of deviation. Such is the art of maneuvering.

23. The Book of Army Management says: On the field of battle, the spoken word does not carry far enough: hence the institution of gongs and drums. Nor can ordinary objects be seen clearly enough: hence the institution of banners and flags.

24. Gongs and drums, banners and flags, are means whereby the ears and eyes of the host may be focused on one particular point.

25. The host thus forming a single united body, is it impossible either for the brave to advance alone, or for the cowardly to retreat alone. This is the art of handling large masses of men.

26. In night-fighting, then, make much use of signal-fires and drums, and in fighting by day, of flags and banners, as a means of influencing the ears and eyes of your army.

27. A whole army may be robbed of its spirit; a commander-in-chief may be robbed of his presence of mind.

28. Now a soldier's spirit is keenest in the morning; by noonday it has begun to flag; and in the evening, his mind is bent only on returning to camp.

29. A clever general, therefore, avoids an army when its spirit is keen, but attacks it when it is sluggish and inclined to return. This is the art of studying moods.

30. Disciplined and calm, to await the appearance of disorder and hubbub amongst the enemy:- this is the art of retaining self-possession.

31. To be near the goal while the enemy is still far from it, to wait at ease while the enemy is toiling and struggling, to be well-fed while the enemy is famished:- this is the art of husbanding one's strength.

32. To refrain from intercepting an enemy whose banners are in perfect order, to refrain from attacking an army drawn up in calm and confident array:- this is the art of studying circumstances.

33. It is a military axiom not to advance uphill against the enemy, nor to oppose him when he comes downhill.

34. Do not pursue an enemy who simulates flight; do not attack soldiers whose temper is keen.

35. Do not swallow bait offered by the enemy. Do not interfere with an army that is returning home.

36. When you surround an army, leave an outlet free. Do not press a desperate foe too hard.

37. Such is the art of warfare.

8
VARIATION IN TACTICS

1. Sun Tzu said: In war, the general receives his commands from the sovereign, collects his army and concentrates his forces

2. When in difficult country, do not encamp. In country where high roads intersect, join hands with your allies. Do not linger in dangerously isolated positions. In hemmed-in situations, you must resort to stratagem. In desperate position, you must fight.

3. There are roads which must not be followed, armies which must be not attacked, towns which must be besieged, positions which must not be contested, commands of the sovereign which must not be obeyed.

4. The general who thoroughly understands the advantages that accompany variation of tactics knows how to handle his troops.

5. The general who does not understand these, may be well acquainted with the configuration of the country, yet he will not be able to turn his knowledge to practical account.

6. So, the student of war who is unversed in the art of war of varying his plans, even though he be acquainted with the Five Advantages, will fail to make the best use of his men.

7. Hence in the wise leader's plans, considerations of advantage and of disadvantage will be blended together.

8. If our expectation of advantage be tempered in this way, we may succeed in accomplishing the essential part of our schemes.

9. If, on the other hand, in the midst of difficulties we are always ready to seize an advantage, we may extricate ourselves from misfortune.

10. Reduce the hostile chiefs by inflicting damage on them; and make trouble for them, and keep them constantly engaged; hold out specious allurements, and make them rush to any given point.

11. The art of war teaches us to rely not on the likelihood of the enemy's not coming, but on our own readiness to receive him; not on the chance of his not attacking, but rather on the fact that we have made our position unassailable.

12. There are five dangerous faults which may affect a general: (1) Recklessness, which leads to destruction; (2) cowardice, which leads to capture; (3) a hasty temper, which can be provoked by insults; (4) a delicacy of honor which is sensitive to shame; (5) over-solicitude for his men, which exposes him to worry and trouble.

13. These are the five besetting sins of a general, ruinous to the conduct of war.

14. When an army is overthrown and its leader slain, the cause will surely be found among these five dangerous faults. Let them be a subject of meditation.

9

THE ARMY ON THE MARCH

1. Sun Tzu said: We come now to the question of encamping the army, and observing signs of the enemy. Pass quickly over mountains, and keep in the neighborhood of valleys.

2. Camp in high places, facing the sun. Do not climb heights in order to fight. So much for mountain warfare.

3. After crossing a river, you should get far away from it.

4. When an invading force crosses a river in its onward march, do not advance to meet it in mid-stream. It will be best to let half the army get across, and then deliver your attack.

5. If you are anxious to fight, you should not go to meet the invader near a river which he has to cross.

6. Moor your craft higher up than the enemy, and facing the sun. Do not move up-stream to meet the enemy. So much for river warfare.

7. In crossing salt-marshes, your sole concern should be to get over them quickly, without any delay.

8. If forced to fight in a salt-marsh, you should have water and grass near you, and get your back to a clump of trees. So much for operations in salt-marshes.

9. In dry, level country, take up an easily accessible position with rising ground to your right and on your rear, so that the danger may be in front, and safety lie behind. So much for campaigning in flat country.

10. These are the four useful branches of military knowledge which enabled the Yellow Emperor to vanquish four several sovereigns.

11. All armies prefer high ground to low and sunny places to dark.

12. If you are careful of your men, and camp on hard ground, the army will be free from disease of every kind, and this will spell victory.

13. When you come to a hill or a bank, occupy the sunny side, with the slope on your right rear. Thus you will at once act for the benefit of your soldiers and utilize the natural advantages of the ground.

14. When, in consequence of heavy rains up-country, a river which you wish to ford is swollen and flecked with foam, you must wait until it subsides.

15. Country in which there are precipitous cliffs with torrents running between, deep natural hollows, confined places, tangled thickets, quagmires and crevasses, should be left with all possible speed and not approached.

16. While we keep away from such places, we should get the enemy to approach them; while we face them, we should let the enemy have them on his rear.

17. If in the neighborhood of your camp there should be any hilly country, ponds surrounded by aquatic grass, hollow basins filled with reeds, or woods with thick undergrowth, they must be carefully routed out and searched; for these are places where men in ambush or insidious spies are likely to be lurking.

18. When the enemy is close at hand and remains quiet, he is relying on the natural strength of his position.

19. When he keeps aloof and tries to provoke a battle, he is anxious for the other side to advance.

20. If his place of encampment is easy of access, he is tendering a bait.

21. Movement amongst the trees of a forest shows that the enemy is advancing. The appearance of a number of screens in the midst of thick grass means that the enemy wants to make us suspicious.

22. The rising of birds in their flight is the sign of an ambuscade. Startled beasts indicate that a sudden attack is coming.

23. When there is dust rising in a high column, it is the sign of chariots advancing; when the dust is low, but spread over a wide area, it betokens the approach of infantry. When it branches out in different directions, it shows that parties have been sent to collect firewood. A few clouds of dust moving to and fro signify that the army is encamping.

24. Humble words and increased preparations are signs that the enemy is about to advance. Violent language and driving forward as if to the attack are signs that he will retreat.

25. When the light chariots come out first and take up a position on the wings, it is a sign that the enemy is forming for battle.

26. Peace proposals unaccompanied by a sworn covenant indicate a plot.

27. When there is much running about and the soldiers fall into rank, it means that the critical moment has come.

28. When some are seen advancing and some retreating, it is a lure.

29. When the soldiers stand leaning on their spears, they are faint from want of food.

30. If those who are sent to draw water begin by drinking themselves, the army is suffering from thirst.

31. If the enemy sees an advantage to be gained and makes no effort to secure it, the soldiers are exhausted.

32. If birds gather on any spot, it is unoccupied. Clamor by night betokens nervousness.

33. If there is disturbance in the camp, the general's authority is weak. If the banners and flags are shifted about, sedition is afoot. If the officers are angry, it means that the men are weary.

34. When an army feeds its horses with grain and kills its cattle for food, and when the men do not hang their cooking-pots over the camp-fires, showing that they will not return to their tents, you may know that they are determined to fight to the death.

35. The sight of men whispering together in small knots or speaking in subdued tones points to disaffection amongst the rank and file.

36. Too frequent rewards signify that the enemy is at the end of his resources; too many punishments betray a condition of dire distress.

37. To begin by bluster, but afterwards to take fright at the enemy's numbers, shows a supreme lack of intelligence.

38. When envoys are sent with compliments in their mouths, it is a sign that the enemy wishes for a truce.

39. If the enemy's troops march up angrily and remain facing ours for a long time without either joining battle or taking themselves off again, the situation is one that demands great vigilance and circumspection.

40. If our troops are no more in number than the enemy, that is amply sufficient; it only means that no direct attack can be made. What we can do is simply to concentrate all our available strength, keep a close watch on the enemy, and obtain reinforcements.

41. He who exercises no forethought but makes light of his opponents is sure to be captured by them.

42. If soldiers are punished before they have grown attached to you, they will not prove submissive; and, unless submissive, then will be practically useless. If, when the soldiers have become attached to you, punishments are not enforced, they will still be unless.

43. Therefore soldiers must be treated in the first instance with humanity, but kept under control by means of iron discipline. This is a certain road to victory.

44. If in training soldiers commands are habitually enforced, the army will be well-disciplined; if not, its discipline will be bad.

45. If a general shows confidence in his men but always insists on his orders being obeyed, the gain will be mutual.

10

TERRAIN

1. Sun Tzu said: We may distinguish six kinds of terrain, to wit: (1) Accessible ground; (2) entangling ground; (3) temporizing ground; (4) narrow passes; (5) precipitous heights; (6) positions at a great distance from the enemy.

2. Ground which can be freely traversed by both sides is called accessible.

3. With regard to ground of this nature, be before the enemy in occupying the raised and sunny spots, and carefully guard your line of supplies. Then you will be able to fight with advantage.

4. Ground which can be abandoned but is hard to re-occupy is called entangling.

5. From a position of this sort, if the enemy is unprepared, you may sally forth and defeat him. But if the enemy is prepared for your coming, and you fail to defeat him, then, return being impossible, disaster will ensue.

6. When the position is such that neither side will gain by making the first move, it is called temporizing ground.

7. In a position of this sort, even though the enemy should offer us an attractive bait, it will be advisable not to stir forth, but rather to retreat, thus

enticing the enemy in his turn; then, when part of his army has come out, we may deliver our attack with advantage.

8. With regard to narrow passes, if you can occupy them first, let them be strongly garrisoned and await the advent of the enemy.

9. Should the army forestall you in occupying a pass, do not go after him if the pass is fully garrisoned, but only if it is weakly garrisoned.

10. With regard to precipitous heights, if you are beforehand with your adversary, you should occupy the raised and sunny spots, and there wait for him to come up.

11. If the enemy has occupied them before you, do not follow him, but retreat and try to entice him away.

12. If you are situated at a great distance from the enemy, and the strength of the two armies is equal, it is not easy to provoke a battle, and fighting will be to your disadvantage.

13. These six are the principles connected with Earth. The general who has attained a responsible post must be careful to study them.

14. Now an army is exposed to six several calamities, not arising from natural causes, but from faults for which the general is responsible. These are: (1) Flight; (2) Insubordination; (3) Collapse; (4) Ruin; (5) Disorganization; (6) Rout.

15. Other conditions being equal, if one force is hurled against another ten times its size, the result will be the Flight of the former.

16. When the common soldiers are too strong and their officers too weak, the result is Insubordination. When the officers are too strong and the common soldiers too weak, the result is Collapse.

17. When the higher officers are angry and insubordinate, and on meeting the enemy give battle on their own account from a feeling of resentment, before the commander-in-chief can tell whether or no he is in a position to fight, the result is ruin.

18. When the general is weak and without authority; when his orders are not clear and distinct; when there are no fixed duties assigned to officers and men, and the ranks are formed in a slovenly haphazard manner, the result is utter Disorganization.

19. When a general, unable to estimate the enemy's strength, allows an inferior force to engage a larger one, or hurls a weak detachment against a powerful one, and neglects to place picked soldiers in the front rank, the result must be Rout.

20. These are six ways of courting defeat, which must be carefully noted by the general who has attained a responsible post.

21. The natural formation of the country is the soldier's best ally; but a power of estimating the adversary, of controlling the forces of victory, and of shrewdly calculating difficulties, dangers and distances, constitutes the test of a great general.

22. He who knows these things, and in fighting puts his knowledge into practice, will win his battles. He who knows them not, nor practices them, will surely be defeated.

23. If fighting is sure to result in victory, then you must fight, even though the ruler forbid it; if fighting will not result in victory, then you must not fight even at the ruler's bidding.

24. The general who advances without coveting fame and retreats without fearing disgrace, whose only thought is to protect his country and do good service for his sovereign, is the jewel of the kingdom.

25. Regard your soldiers as your children, and they will follow you into the deepest valleys; look upon them as your own beloved sons, and they will stand by you even unto death.

26. If, however, you are indulgent, but unable to make your authority felt; kind-hearted, but unable to enforce your commands; and incapable, moreover, of quelling disorder: then your soldiers must be likened to spoilt children; they are useless for any practical purpose.

27. If we know that our own men are in a condition to attack, but are unaware that the enemy is not open to attack, we have gone only halfway towards victory.

28. If we know that the enemy is open to attack, but are unaware that our own men are not in a condition to attack, we have gone only halfway towards victory.

29. If we know that the enemy is open to attack, and also know that our men are in a condition to attack, but are unaware that the nature of the ground makes fighting impracticable, we have still gone only halfway towards victory.

30. Hence the experienced soldier, once in motion, is never bewildered; once he has broken camp, he is never at a loss.

31. Hence the saying: If you know the enemy and know yourself, your victory will not stand in doubt; if you know Heaven and know Earth, you may make your victory complete.

11
THE NINE SITUATIONS

1. Sun Tzu said: The art of war recognizes nine varieties of ground: (1) Dispersive ground; (2) facile ground; (3) contentious ground; (4) open ground; (5) ground of intersecting highways; (6) serious ground; (7) difficult ground; (8) hemmed-in ground; (9) desperate ground.

2. When a chieftain is fighting in his own territory, it is dispersive ground.

3. When he has penetrated into hostile territory, but to no great distance, it is facile ground.

4. Ground the possession of which imports great advantage to either side, is contentious ground.

5. Ground on which each side has liberty of movement is open ground.

6. Ground which forms the key to three contiguous states, so that he who occupies it first has most of the Empire at his command, is a ground of intersecting highways.

7. When an army has penetrated into the heart of a hostile country, leaving a number of fortified cities in its rear, it is serious ground.

8. Mountain forests, rugged steeps, marshes and fens--all country that is hard to traverse: this is difficult ground.

9. Ground which is reached through narrow gorges, and from which we can only retire by tortuous paths, so that a small number of the enemy would suffice to crush a large body of our men: this is hemmed in ground.

10. Ground on which we can only be saved from destruction by fighting without delay, is desperate ground.

11. On dispersive ground, therefore, fight not. On facile ground, halt not. On contentious ground, attack not.

12. On open ground, do not try to block the enemy's way. On the ground of intersecting highways, join hands with your allies.

13. On serious ground, gather in plunder. In difficult ground, keep steadily on the march.

14. On hemmed-in ground, resort to stratagem. On desperate ground, fight.

15. Those who were called skillful leaders of old knew how to drive a wedge between the enemy's front and rear; to prevent co-operation between his large and small divisions; to hinder the good troops from rescuing the bad, the officers from rallying their men.

16. When the enemy's men were united, they managed to keep them in disorder.

17. When it was to their advantage, they made a forward move; when otherwise, they stopped still.

18. If asked how to cope with a great host of the enemy in orderly array and on the point of marching to the attack, I should say: "Begin by seizing something which your opponent holds dear; then he will be amenable to your will."

19. Rapidity is the essence of war: take advantage of the enemy's unreadiness, make your way by unexpected routes, and attack unguarded spots.

20. The following are the principles to be observed by an invading force: The further you penetrate into a country, the greater will be the solidarity of your troops, and thus the defenders will not prevail against you.

21. Make forays in fertile country in order to supply your army with food.

22. Carefully study the well-being of your men, and do not overtax them. Concentrate your energy and hoard your strength. Keep your army continually on the move, and devise unfathomable plans.

23. Throw your soldiers into positions whence there is no escape, and they will prefer death to flight. If they will face death, there is nothing they may not achieve. Officers and men alike will put forth their uttermost strength.

24. Soldiers when in desperate straits lose the sense of fear. If there is no place of refuge, they will stand firm. If they are in hostile country, they will show a stubborn front. If there is no help for it, they will fight hard.

25. Thus, without waiting to be marshaled, the soldiers will be constantly on the qui vive; without waiting to be asked, they will do your will; without restrictions, they will be faithful; without giving orders, they can be trusted.

26. Prohibit the taking of omens, and do away with superstitious doubts. Then, until death itself comes, no calamity need be feared.

27. If our soldiers are not overburdened with money, it is not because they have a distaste for riches; if their lives are not unduly long, it is not because they are disinclined to longevity.

28. On the day they are ordered out to battle, your soldiers may weep, those sitting up bedewing their garments, and those lying down letting the tears

run down their cheeks. But let them once be brought to bay, and they will display the courage of a Chu or a Kuei.

29. The skillful tactician may be likened to the Shuai-Jan. Now the Shuai-Jan is a snake that is found in the Ch`ang mountains. Strike at its head, and you will be attacked by its tail; strike at its tail, and you will be attacked by its head; strike at its middle, and you will be attacked by head and tail both.

30. Asked if an army can be made to imitate the Shuai-Jan, I should answer, Yes. For the men of Wu and the men of Yueh are enemies; yet if they are crossing a river in the same boat and are caught by a storm, they will come to each other's assistance just as the left hand helps the right.

31. Hence it is not enough to put one's trust in the tethering of horses, and the burying of chariot wheels in the ground

32. The principle on which to manage an army is to set up one standard of courage which all must reach.

33. How to make the best of both strong and weak--that is a question involving the proper use of ground.

34. Thus the skillful general conducts his army just as though he were leading a single man, willy-nilly, by the hand.

35. It is the business of a general to be quiet and thus ensure secrecy; upright and just, and thus maintain order.

36. He must be able to mystify his officers and men by false reports and appearances, and thus keep them in total ignorance.

37. By altering his arrangements and changing his plans, he keeps the enemy without definite knowledge. By shifting his camp and taking circuitous routes, he prevents the enemy from anticipating his purpose.

38. At the critical moment, the leader of an army acts like one who has climbed up a height and then kicks away the ladder behind him. He carries his men deep into hostile territory before he shows his hand.

39. He burns his boats and breaks his cooking-pots; like a shepherd driving a flock of sheep, he drives his men this way and that, and nothing knows whither he is going.

40. To muster his host and bring it into danger:- this may be termed the business of the general.

41. The different measures suited to the nine varieties of ground; the expediency of aggressive or defensive tactics; and the fundamental laws of human nature: these are things that must most certainly be studied.

42. When invading hostile territory, the general principle is, that penetrating deeply brings cohesion; penetrating but a short way means dispersion.

43. When you leave your own country behind, and take your army across neighborhood territory, you find yourself on critical ground. When there are means of communication on all four sides, the ground is one of intersecting highways.

44. When you penetrate deeply into a country, it is serious ground. When you penetrate but a little way, it is facile ground.

45. When you have the enemy's strongholds on your rear, and narrow passes in front, it is hemmed-in ground. When there is no place of refuge at all, it is desperate ground.

46. Therefore, on dispersive ground, I would inspire my men with unity of purpose. On facile ground, I would see that there is close connection between all parts of my army.

47. On contentious ground, I would hurry up my rear.

48. On open ground, I would keep a vigilant eye on my defenses. On ground of intersecting highways, I would consolidate my alliances.

49. On serious ground, I would try to ensure a continuous stream of supplies. On difficult ground, I would keep pushing on along the road.

50. On hemmed-in ground, I would block any way of retreat. On desperate ground, I would proclaim to my soldiers the hopelessness of saving their lives.

51. For it is the soldier's disposition to offer an obstinate resistance when surrounded, to fight hard when he cannot help himself, and to obey promptly when he has fallen into danger.

52. We cannot enter into alliance with neighboring princes until we are acquainted with their designs. We are not fit to lead an army on the march unless we are familiar with the face of the country - its mountains and forests, its pitfalls and precipices, its marshes and swamps. We shall be unable to turn natural advantages to account unless we make use of local guides.

53. To be ignored of any one of the following four or five principles does not befit a warlike prince.

54. When a warlike prince attacks a powerful state, his generalship shows itself in preventing the concentration of the enemy's forces. He overawes his opponents, and their allies are prevented from joining against him.

55. Hence he does not strive to ally himself with all and sundry, nor does he foster the power of other states. He carries out his own secret designs, keeping his antagonists in awe. Thus he is able to capture their cities and overthrow their kingdoms.

56. Bestow rewards without regard to rule, issue orders without regard to previous arrangements; and you will be able to handle a whole army as though you had to do with but a single man.

57. Confront your soldiers with the deed itself; never let them know your design. When the outlook is bright, bring it before their eyes; but tell them nothing when the situation is gloomy.

58. Place your army in deadly peril, and it will survive; plunge it into desperate straits, and it will come off in safety.

59. For it is precisely when a force has fallen into harm's way that is capable of striking a blow for victory.

60. Success in warfare is gained by carefully accommodating ourselves to the enemy's purpose.

61. By persistently hanging on the enemy's flank, we shall succeed in the long run in killing the commander-in-chief.

62. This is called ability to accomplish a thing by sheer cunning.

63. On the day that you take up your command, block the frontier passes, destroy the official tallies, and stop the passage of all emissaries.

64. Be stern in the council-chamber, so that you may control the situation.

65. If the enemy leaves a door open, you must rush in.

66. Forestall your opponent by seizing what he holds dear, and subtly contrive to time his arrival on the ground.

67. Walk in the path defined by rule, and accommodate yourself to the enemy until you can fight a decisive battle.

68. At first, then, exhibit the coyness of a maiden, until the enemy gives you an opening; afterwards emulate the rapidity of a running hare, and it will be too late for the enemy to oppose you.

12

THE ATTACK BY FIRE

1. Sun Tzu said: There are five ways of attacking with fire. The first is to burn soldiers in their camp; the second is to burn stores; the third is to burn baggage trains; the fourth is to burn arsenals and magazines; the fifth is to hurl dropping fire amongst the enemy.

2. In order to carry out an attack, we must have means available. The material for raising fire should always be kept in readiness.

3. There is a proper season for making attacks with fire, and special days for starting a conflagration.

4. The proper season is when the weather is very dry; the special days are those when the moon is in the constellations of the Sieve, the Wall, the Wing or the Cross-bar; for these four are all days of rising wind.

5. In attacking with fire, one should be prepared to meet five possible developments:

6. (1) When fire breaks out inside to enemy's camp, respond at once with an attack from without.

7. (2) If there is an outbreak of fire, but the enemy's soldiers remain quiet, bide your time and do not attack.

8. (3) When the force of the flames has reached its height, follow it up with an attack, if that is practicable; if not, stay where you are.

9. (4) If it is possible to make an assault with fire from without, do not wait for it to break out within, but deliver your attack at a favorable moment.

10. (5) When you start a fire, be to windward of it. Do not attack from the leeward.

11. A wind that rises in the daytime lasts long, but a night breeze soon falls.

12. In every army, the five developments connected with fire must be known, the movements of the stars calculated, and a watch kept for the proper days.

13. Hence those who use fire as an aid to the attack show intelligence; those who use water as an aid to the attack gain an accession of strength.

14. By means of water, an enemy may be intercepted, but not robbed of all his belongings.

15. Unhappy is the fate of one who tries to win his battles and succeed in his attacks without cultivating the spirit of enterprise; for the result is waste of time and general stagnation.

16. Hence the saying: The enlightened ruler lays his plans well ahead; the good general cultivates his resources.

17. Move not unless you see an advantage; use not your troops unless there is something to be gained; fight not unless the position is critical.

18. No ruler should put troops into the field merely to gratify his own spleen; no general should fight a battle simply out of pique.

19. If it is to your advantage, make a forward move; if not, stay where you are.

20. Anger may in time change to gladness; vexation may be succeeded by content.

21. But a kingdom that has once been destroyed can never come again into being; nor can the dead ever be brought back to life.

22. Hence the enlightened ruler is heedful, and the good general full of caution. This is the way to keep a country at peace and an army intact.

13
THE USE OF SPIES

1. Sun Tzu said: Raising a host of a hundred thousand men and marching them great distances entails heavy loss on the people and a drain on the resources of the State. The daily expenditure will amount to a thousand ounces of silver. There will be commotion at home and abroad, and men will drop down exhausted on the highways. As many as seven hundred thousand families will be impeded in their labor.

2. Hostile armies may face each other for years, striving for the victory which is decided in a single day. This being so, to remain in ignorance of the enemy's condition simply because one grudges the outlay of a hundred ounces of silver in honors and emoluments, is the height of inhumanity.

3. One who acts thus is no leader of men, no present help to his sovereign, no master of victory.

4. Thus, what enables the wise sovereign and the good general to strike and conquer, and achieve things beyond the reach of ordinary men, is foreknowledge.

5. Now this foreknowledge cannot be elicited from spirits; it cannot be obtained inductively from experience, nor by any deductive calculation.

6. Knowledge of the enemy's dispositions can only be obtained from other men.

221

7. Hence the use of spies, of whom there are five classes: (1) Local spies; (2) inward spies; (3) converted spies; (4) doomed spies; (5) surviving spies.

8. When these five kinds of spy are all at work, none can discover the secret system. This is called "divine manipulation of the threads." It is the sovereign's most precious faculty.

9. Having local spies means employing the services of the inhabitants of a district.

10. Having inward spies, making use of officials of the enemy.

11. Having converted spies, getting hold of the enemy's spies and using them for our own purposes.

12. Having doomed spies, doing certain things openly for purposes of deception, and allowing our spies to know of them and report them to the enemy.

13. Surviving spies, finally, are those who bring back news from the enemy's camp.

14. Hence it is that which none in the whole army are more intimate relations to be maintained than with spies. None should be more liberally rewarded. In no other business should greater secrecy be preserved.

15. Spies cannot be usefully employed without a certain intuitive sagacity.

16. They cannot be properly managed without benevolence and straightforwardness.

17. Without subtle ingenuity of mind, one cannot make certain of the truth of their reports.

18. Be subtle! be subtle! and use your spies for every kind of business.

19. If a secret piece of news is divulged by a spy before the time is ripe, he must be put to death together with the man to whom the secret was told.

20. Whether the object be to crush an army, to storm a city, or to assassinate an individual, it is always necessary to begin by finding out the names of the attendants, the aides-de-camp, and door-keepers and sentries of the general in command. Our spies must be commissioned to ascertain these.

21. The enemy's spies who have come to spy on us must be sought out, tempted with bribes, led away and comfortably housed. Thus they will become converted spies and available for our service.

22. It is through the information brought by the converted spy that we are able to acquire and employ local and inward spies.

23. It is owing to his information, again, that we can cause the doomed spy to carry false tidings to the enemy.

24. Lastly, it is by his information that the surviving spy can be used on appointed occasions.

25. The end and aim of spying in all its five varieties is knowledge of the enemy; and this knowledge can only be derived, in the first instance, from the converted spy. Hence it is essential that the converted spy be treated with the utmost liberality.

26. Of old, the rise of the Yin dynasty was due to I Chih who had served under the Hsia. Likewise, the rise of the Chou dynasty was due to Lu Ya who had served under the Yin.

27. Hence it is only the enlightened ruler and the wise general who will use the highest intelligence of the army for purposes of spying and thereby they achieve great results. Spies are a most important element in war, because on them depends an army's ability to move.

Practice Answer Sheet

1. Laying Plans

Q1.   (a)   (b)   (c)   (d)

Q2.   (a)   (b)   (c)   (d)

Q3.   (a)   (b)   (c)   (d)

Q4.   (a)   (b)   (c)   (d)

Q5.   (a)   (b)   (c)   (d)

Q6.   (a)   (b)   (c)   (d)

Q7.   (a)   (b)   (c)   (d)

Q8.   (a)   (b)   (c)   (d)

Q9.   (a)   (b)   (c)   (d)

Q10.   (a)   (b)   (c)   (d)

Q11.   (a)   (b)   (c)   (d)

Q12.   (a)   (b)   (c)   (d)

Q13.   (a)   (b)   (c)   (d)

Q14.   (a)   (b)   (c)   (d)

Q15.   (a)   (b)   (c)   (d)

Q16.   (a)   (b)   (c)   (d)

Practice Answer Sheet

2. Waging Wars

Q1.	(a)	(b)	(c)	(d)
Q2.	(a)	(b)	(c)	(d)
Q3.	True	False		
Q4.	True	False		
Q5.	(a)	(b)	(c)	(d)
Q6.	True	False		
Q7.	True	False		
Q8.	(a)	(b)	(c)	(d)
Q9.	(a)	(b)	(c)	(d)
Q10.	(a)	(b)	(c)	(d)
Q11.	True	False		
Q12.	(a)	(b)	(c)	(d)

Practice Answer Sheet

3. Attack by Stratagem

Q1. (a) (b) (c) (d)
Q2. (a) (b) (c) (d)
Q3. (a) (b) (c) (d)
Q4. (a) (b) (c) (d)
Q5. (a) (b) (c) (d)
Q6. (a) (b) (c) (d)
Q7. True False
Q8. (a) (b) (c) (d)
Q9. (a) (b) (c) (d)
Q10. (a) (b) (c) (d)
Q11. (a) (b) (c) (d)
Q12. (a) (b) (c) (d)
Q13. (a) (b) (c) (d)
Q14. (a) (b) (c) (d)
Q15. (a) (b) (c) (d)

Practice Answer Sheet

4. Tactical Disposition

Q1.   (a)   (b)   (c)   (d)
Q2.   (a)   (b)   (c)   (d)
Q3.   True   False
Q4.   (a)   (b)   (c)   (d)
Q5.   (a)   (b)   (c)   (d)
Q6.   True   False
Q7.   True   Flase
Q8.   (a)   (b)   (c)   (d)
Q9.   (a)   (b)   (c)   (d)
Q10. (a)   (b)   (c)   (d)
Q11. (a)   (b)   (c)   (d)
Q12. (a)   (b)   (c)   (d)

5. Energy

Q1.   (a)   (b)   (c)   (d)

Q2.   (a)   (b)   (c)   (d)

Q3.   (a)   (b)   (c)   (d)

Q4.   (a)   (b)   (c)   (d)

Q5.   (a)   (b)   (c)   (d)

Q6.   (a)   (b)   (c)   (d)

Q7.   (a)   (b)   (c)   (d)

Q8.   True   False

Q9.   (a)   (b)   (c)   (d)

Q10.   (a)   (b)   (c)   (d)

Q11.   (a)   (b)   (c)   (d)

Q12.   True   False

Q13.   (a)   (b)   (c)   (d)

Q14.   (a)   (b)   (c)   (d)

Q15.   (a)   (b)   (c)   (d)

Practice Answer Sheet

6. Weak Points and Strong

Q1.   (a)   (b)   (c)   (d)
Q2.   True   False
Q3.   (a)   (b)   (c)   (d)
Q4.   (a)   (b)   (c)   (d)
Q5.   True   False
Q6.   (a)   (b)   (c)   (d)
Q7.   (a)   (b)   (c)   (d)
Q8.   (a)   (b)   (c)   (d)
Q9.   (a)   (b)   (c)   (d)
Q10. (a)   (b)   (c)   (d)
Q11. (a)   (b)   (c)   (d)
Q12. (a)   (b)   (c)   (d)
Q13. True   False
Q14. (a)   (b)   (c)   (d)
Q15. True   False
Q16. (a)   (b)   (c)   (d)
Q17. (a)   (b)   (c)   (d)
Q18. (a)   (b)   (c)   (d)
Q19. (a)   (b)   (c)   (d)
Q20. (a)   (b)   (c)   (d)
Q21. True   False
Q22. (a)   (b)   (c)   (d)
Q23. (a)   (b)   (c)   (d)
Q24. (a)   (b)   (c)   (d)

Practice Answer Sheet

7. Maneuvering

Q1.   (a)   (b)   (c)   (d)
Q2.   (a)   (b)   (c)   (d)
Q3.   (a)   (b)   (c)   (d)
Q4.   (a)   (b)   (c)   (d)
Q5.   (a)   (b)   (c)   (d)
Q6.   (a)   (b)   (c)   (d)
Q7.   (a)   (b)   (c)   (d)
Q8.   True   False
Q9.   True   False
Q10.   True   False
Q11.   (a)   (b)   (c)   (d)
Q12.   (a)   (b)   (c)   (d)
Q13.   True   False
Q14.   (a)   (b)   (c)   (d)
Q15.   (a)   (b)   (c)   (d)
Q16.   (a)   (b)   (c)   (d)
Q17.   (a)   (b)   (c)   (d)
Q18.   True   False
Q19.   (a)   (b)   (c)   (d)
Q20.   (a)   (b)   (c)   (d)
Q21.   (a)   (b)   (c)   (d)
Q22.   (a)   (b)   (c)   (d)
Q23.   (a)   (b)   (c)   (d)
Q24.   (a)   (b)   (c)   (d)
Q25.   (a)   (b)   (c)   (d)
Q26.   (a)   (b)   (c)   (d)

Practice Answer Sheet

8. Variation in Tactics

Q1.	True	False		
Q2.	(a)	(b)	(c)	(d)
Q3.	(a)	(b)	(c)	(d)
Q4.	(a)	(b)	(c)	(d)
Q5.	True	False		
Q6.	True	False		
Q7.	(a)	(b)	(c)	(d)
Q8.	(a)	(b)	(c)	(d)
Q9.	(a)	(b)	(c)	(d)
Q10.	(a)	(b)	(c)	(d)

Practice Answer Sheet

9. The Army on the March

Q1.	(a)	(b)	(c)	(d)
Q2.	(a)	(b)	(c)	(d)
Q3.	(a)	(b)	(c)	(d)
Q4.	(a)	(b)	(c)	(d)
Q5.	(a)	(b)	(c)	(d)
Q6.	(a)	(b)	(c)	(d)
Q7.	(a)	(b)	(c)	(d)
Q8.	(a)	(b)	(c)	(d)
Q9.	True	False		
Q10.	(a)	(b)	(c)	(d)
Q11.	(a)	(b)	(c)	(d)
Q12.	(a)	(b)	(c)	(d)
Q13.	(a)	(b)	(c)	(d)
Q14.	(a)	(b)	(c)	(d)
Q15.	True	False		
Q16.	(a)	(b)	(c)	(d)
Q17.	True	False		
Q18.	True	False		
Q19.	True	False		
Q20.	True	False		
Q21.	True	False		
Q22.	(a)	(b)	(c)	(d)
Q23.	True	False		
Q24.	(a)	(b)	(c)	(d)
Q25.	(a)	(b)	(c)	(d)
Q26.	(a)	(b)	(c)	(d)

Q27.	(a)	(b)	(c)	(d)
Q28.	(a)	(b)	(c)	(d)
Q29.	(a)	(b)	(c)	(d)
Q30.	(a)	(b)	(c)	(d)
Q31.	(a)	(b)	(c)	(d)

Practice Answer Sheet

10. Terrain

Q1.   (a)   (b)   (c)   (d)
Q2.   (a)   (b)   (c)   (d)
Q3.   (a)   (b)   (c)   (d)
Q4.   (a)   (b)   (c)   (d)
Q5.   (a)   (b)   (c)   (d)
Q6.   (a)   (b)   (c)   (d)
Q7.   (a)   (b)   (c)   (d)
Q8.   (a)   (b)   (c)   (d)
Q9.   (a)   (b)   (c)   (d)
Q10. True   False
Q11. (a)   (b)   (c)   (d)
Q12. True   False
Q13. (a)   (b)   (c)   (d)
Q14. (a)   (b)   (c)   (d)
Q15. (a)   (b)   (c)   (d)
Q16. (a)   (b)   (c)   (d)
Q17. (a)   (b)   (c)   (d)
Q18. (a)   (b)   (c)   (d)
Q19. (a)   (b)   (c)   (d)
Q20. True   False
Q21. (a)   (b)   (c)   (d)
Q22. (a)   (b)   (c)   (d)
Q23. (a)   (b)   (c)   (d)
Q24. (a)   (b)   (c)   (d)
Q25. (a)   (b)   (c)   (d)

Practice Answer Sheet

11. The Nine Situations

Q1.	(a)	(b)	(c)	(d)
Q2.	(a)	(b)	(c)	(d)
Q3.	(a)	(b)	(c)	(d)
Q4.	(a)	(b)	(c)	(d)
Q5.	(a)	(b)	(c)	(d)
Q6.	(a)	(b)	(c)	(d)
Q7.	(a)	(b)	(c)	(d)
Q8.	(a)	(b)	(c)	(d)
Q9.	(a)	(b)	(c)	(d)
Q10.	(a)	(b)	(c)	(d)
Q11.	(a)	(b)	(c)	(d)
Q12.	(a)	(b)	(c)	(d)
Q13.	(a)	(b)	(c)	(d)
Q14.	(a)	(b)	(c)	(d)
Q15.	(a)	(b)	(c)	(d)
Q16.	(a)	(b)	(c)	(d)
Q17.	(a)	(b)	(c)	(d)
Q18.	True	False		
Q19.	(a)	(b)	(c)	(d)
Q20.	(a)	(b)	(c)	(d)
Q21.	(a)	(b)	(c)	(d)
Q22.	(a)	(b)	(c)	(d)
Q23.	(a)	(b)	(c)	(d)
Q24.	True	False		
Q25.	True	False		
Q26.	True	False		

Q27.	(a)	(b)	(c)	(d)
Q28.	(a)	(b)	(c)	(d)
Q29.	(a)	(b)	(c)	(d)
Q30.	(a)	(b)	(c)	(d)
Q31.	(a)	(b)	(c)	(d)
Q32.	True	False		
Q33.	(a)	(b)	(c)	(d)

Practice Answer Sheet

12. The Attack by Fire

Q1.   (a)   (b)   (c)   (d)
Q2.   (a)   (b)   (c)   (d)
Q3.   (a)   (b)   (c)   (d)
Q4.   (a)   (b)   (c)   (d)
Q5.   (a)   (b)   (c)   (d)
Q6.   (a)   (b)   (c)   (d)
Q7.   (a)   (b)   (c)   (d)
Q8.   (a)   (b)   (c)   (d)
Q9.   (a)   (b)   (c)   (d)
Q10.   (a)   (b)   (c)   (d)
Q11.   (a)   (b)   (c)   (d)
Q12.   True   False
Q13.   (a)   (b)   (c)   (d)
Q14.   (a)   (b)   (c)   (d)
Q15.   (a)   (b)   (c)   (d)
Q16.   (a)   (b)   (c)   (d)
Q17.   (a)   (b)   (c)   (d)

Practice Answer Sheet

13. The Use of Spies

Q1.   (a)   (b)   (c)   (d)
Q2.   (a)   (b)   (c)   (d)
Q3.   (a)   (b)   (c)   (d)
Q4.   (a)   (b)   (c)   (d)
Q5.   True   False
Q6.   (a)   (b)   (c)   (d)
Q7.   (a)   (b)   (c)   (d)
Q8.   (a)   (b)   (c)   (d)
Q9.   (a)   (b)   (c)   (d)
Q10. (a)   (b)   (c)   (d)
Q11. (a)   (b)   (c)   (d)
Q12. (a)   (b)   (c)   (d)
Q13. (a)   (b)   (c)   (d)
Q14. (a)   (b)   (c)   (d)
Q15. (a)   (b)   (c)   (d)
Q16. (a)   (b)   (c)   (d)
Q17. (a)   (b)   (c)   (d)